TOURISM AND CULTURAL CHANGE: 40

Explorer Travellers and Adventure Tourism

Jennifer Laing and Warwick Frost

CHANNEL VIEW PUBLICATIONS
Bristol • Buffalo • Toronto

Library of Congress Cataloging in Publication Data
Laing, Jennifer
Explorer Travellers and Adventure Tourism/Jennifer Laing and Warwick Frost.
Tourism and Cultural Change: 40
Includes bibliographical references and index.
1. Adventure travel. 2. Tourism. I. Frost, Warwick. II. Title.
G516.L35 2014
910.4–dc232014008997

British Library Cataloguing in Publication Data
A catalogue entry for this book is available from the British Library.

ISBN-13: 978-1-84541-458-0 (hbk)
ISBN-13: 978-1-84541-457-3 (pbk)

Channel View Publications
UK: St Nicholas House, 31–34 High Street, Bristol BS1 2AW, UK.
USA: UTP, 2250 Military Road, Tonawanda, NY 14150, USA.
Canada: UTP, 5201 Dufferin Street, North York, Ontario M3H 5T8, Canada.

Website: www.channelviewpublications.com
Twitter: Channel_View
Facebook: https://www.facebook.com/channelviewpublications
Blog: www.channelviewpublications.wordpress.com

The policy of Multilingual Matters/Channel View Publications is to use papers that are natural, renewable and recyclable products, made from wood grown in sustainable forests. In the manufacturing process of our books, and to further support our policy, preference is given to printers that have FSC and PEFC Chain of Custody certification. The FSC and/or PEFC logos will appear on those books where full certification has been granted to the printer concerned.

Typeset by Techset Composition India(P) Ltd., Bangalore and Chennai, India.
Printed and bound in Great Britain by Short Run Press Ltd.

Explorer Travellers and
Adventure Tourism

TOURISM AND CULTURAL CHANGE

Series Editors: Professor Mike Robinson, *Ironbridge International Institute for Cultural Heritage, University of Birmingham, UK* and Dr Alison Phipps, *University of Glasgow, Scotland, UK*

TCC is a series of books that explores the complex and ever-changing relationship between tourism and culture(s). The series focuses on the ways that places, peoples, pasts and ways of life are increasingly shaped/transformed/created/packaged for touristic purposes. The series examines the ways tourism utilises/makes and re-makes cultural capital in its various guises (visual and performing arts, crafts, festivals, built heritage, cuisine, etc.) and the multifarious political, economic, social and ethical issues that are raised as a consequence.

Understanding tourism's relationships with culture(s), and vice versa, is of ever-increasing significance in a globalising world. This series will critically examine the dynamic inter-relationships between tourism and culture(s). Theoretical explorations, research-informed analyses, and detailed historical reviews from a variety of disciplinary perspectives are invited to consider such relationships.

Full details of all the books in this series and of all our other publications can be found on http://www.channelviewpublications.com, or by writing to Channel View Publications, St Nicholas House, 31–34 High Street, Bristol BS1 2AW, UK.

Contents

1 Introducing the Explorer Traveller

Prologue: *A Time of Gifts* (Patrick Leigh Fermor, 1977)

In 1933, 18-year-old Patrick Leigh Fermor set out on a big adventure. He was in London, supposedly studying for exams that would get him into the army officer school at Sandhurst. In reality, he was bored, distracted and dispirited. On an overcast autumn afternoon, an idea suddenly sprang into his head:

> Change scenery; abandon London and England and set out across Europe like a tramp – or, as I characteristically phrased it to myself, like a pilgrim or a palmer, an errant scholar, a broken knight … I would travel on foot, sleep in hayricks in summer, shelter in barns when it was raining or snowing and only consort with peasants and tramps. (Fermor, 1977: 20)

His plan was to walk to Constantinople in Turkey, a destination inspired by his love of classical literature and which would allow a detour via Greece (Cooper, 2012). The rough route was to follow first the Rhine and then the Danube. His clothes and equipment were cheaply purchased from an army surplus store. His budget was an allowance of one pound per week. The adventure was solo:

> I wondered during the first few days whether to enlist a companion; but I knew that the enterprise had to be solitary and the break complete. I wanted to think, write, stay or move on at my own speed and unencumbered, to gaze at things with a changed eye and listen to new tongues that were untainted by a single familiar word. (Fermor, 1977: 21)

After an uneventful start in the Netherlands, Fermor approached the German border with trepidation. Born during WWI, for all of his life Germany had been demonised as the enemy. His first taste was the little town of Goch:

The town was hung with National Socialist flags and a window of an outfitter's shop next door held a display of Party equipment: swastika arm-bands, daggers for the Hitler Youth, blouses for Hitler Maidens and brown shirts for grown-up SA men; swastika buttonholes were arranged in a pattern which read *Heil Hitler*. (Fermor, 1977: 43)

Despite this, Fermor found he warmed to the Germans. He could speak a bit of their language and worked hard to improve. Many of them were interested in history, music and literature just like him. Most were very welcoming, for 'there is an old tradition in Germany of benevolence to the wandering young: the very humility of my status acted as an Open Sesame to kindness and hospitality' (Fermor, 1977: 50). An innkeeper invited Fermor to share Christmas with him and his family, and in pubs and cafes many welcomed the chance to talk with a stranger.

Fermor grappled with this contradictory juxtaposition of friendly hospitality and the embrace of National Socialism. In a Munich bar, he made friends with some factory workers about his own age. One offered to put him up for the night. Proudly, his new-found friend showed him his house:

The room turned out to be a shrine of Hitlerania. The walls were covered with flags, photographs, posters, slogans and emblems. His S.A. Uniform hung neatly ironed on a hanger. He explained these cult objects with fetishist zest, saving up to last the centrepiece of his collection. It was an automatic pistol, a Luger. (Fermor, 1977: 129–130)

His friend explained that this was all new; he was only a recent convert:

You should have seen it last year! You would have laughed! Then it was all red flags, stars, hammers and sickles, pictures of Lenin and Stalin ... You should have seen me! Street fights! We used to beat the hell out of the Nazis and they beat the hell out of us ... Then suddenly, when Hitler came into power, I understood it was all nonsense and lies. I realized Adolf was the man for me. All of a sudden! ... They changed too! – all those chaps in the bar. Every single one! They're all in the S.A. now. (Fermor, 1977: 130)

Crossing over into Austria, Fermor entered a new realm. This was the wreckage of the Austro-Hungarian Empire, broken up after WWI. Here, he found that people were looking backwards rather than to the future. There was an elegiac sadness, a despair at what might be coming. Partly, this was a reaction to the changes in Germany, but there was also a more general rejection of modernity. For example, in an inn by the Danube, he fell into conversation with a man in old-fashioned leather breeches. They talked of Wagner, castles, medieval knights and fishing. However, gradually the man became more and more depressed:

> Everything is going to vanish! They talk of building power-dams across the Danube and I tremble whenever I think of it! They'll make the wildest river in Europe as tame as a municipal waterworks. All those fish from the East – they would never come back! Never, never, never! (Fermor, 1977: 161–162)

Introducing the Explorer Traveller

Today, we might see Fermor as representative of many young tourists. He is taking a gap year, backpacking across Europe. He is trying to get off the beaten path and meet some locals (though at the same time he's got a list of famous sights to take in). He is on a tight budget, scrounging beds and meals where he can. He chases girls, trying to impress them with his knowledge of modern music. He writes in a journal, although he loses this when he gets drunk at a pub. It seems a familiar story. Except he is not following a familiar path. In 1933, young people did not travel this way. There were no backpackers, no youth-orientated travel agents or tour operators, no destination marketing strategies aimed at him. There were youth hostels; they started in Germany in 1912, but he only went to one and there his backpack was stolen.

Fermor is a pioneer, a path-finder. He is an example of a small group of elite travellers undertaking adventurous travel. They go where no-one has been before, or at least where they imagine no-one has been. Their accounts – whether through books, television or increasingly the internet and other new media – open up new places and experiences, inspiring others to follow. Whether through reading them directly or picking up parts of their stories second hand, others are stimulated to try and have similar travels and adventures and push their own boundaries.

Our aim in this book is to examine this phenomenon of travel based on the desire for adventure and the need to explore. We are particularly interested in these elite travellers who set out to be first and provide the inspiration for later flows of tourists. Some write accounts that have made them widely known around the world, bywords for adventure-filled travel and read by many. In addition to Fermor, examples of these giants of travel literature include Richard Burton, Henry Morton Stanley, Thor Heyerdahl, Tim Severin, Dervla Murphy, Eric Newby, Edith Durham and Isabella Bird. Even today, with so much of the world seemingly explored and well-trodden, there is a constant flow of new explorers. They strive to climb all the highest peaks on the seven continents, recreate the journeys of ancient adventurers, trek in dangerous seasons, skateboard across the Nullarbor, and freefall for extraordinary distances at seemingly impossible speeds. Others immerse themselves in indigenous communities, even marrying into them. The modern media, whether online, print or television, continues to have a strong fascination with unusual, heroic, even reckless travellers.

We call these adventurers *explorer travellers*. Their identities and exploits are tied in with the myths of the explorer and the discoverer. It is a constructed fantasy that permeates modern tourism. Imagining ourselves to be explorers like them gives purpose and status to our travels. Rather than just being tourists on holiday, we are seeking, searching, discovering. There is a primal urge to keep moving, looking over the next hill for new lands, peoples and adventures. Part of the myth is that we expect the experience will transform us. We will return as a better, wiser, more interesting person. Our travels will be our travails. They will test us and we will see just how far our boundaries extend. Such imaginings push us beyond pleasure-seeking and relaxing holidays. They transport us to a different fantasy world in which we fashion new identities as travellers and explorers emulating real and fictional heroes.

The explorer travellers form an elite which leads the way for the rest of us. Most tourists will not engage in anything like the dangerous activities of this elite, but they are powerfully influenced by their experiences and their mystique. The explorer travellers are the trendsetters, the mavens who shape the fashions of modern travel. Their influence has been powerful since the early 19th century and there is no sign of it diminishing.

In this book, we examine the worlds and influence of explorer travellers through their own first-hand accounts. The need to tell their story to the outside world seems to be one of the defining characteristics of explorer travellers. The bases of this study are their personal stories. Our research draws on two major groups of sources. The first are semi-structured long interviews that we conducted with 39 modern-day explorer travellers from around the globe (these are listed in Part A of our sources). Our second group of sources are published explorer and traveller accounts, mainly factual, but including some fictional novels (see Part B of our sources).

We have chosen the 19th century as the starting point for our analysis. While we recognise that there were influential traveller narratives before then, it is really only in the 19th century that these become a very popular genre, widely read and inspiring others to follow. It is tempting to view the 19th century as a *Golden Age* of exploration, although this needs qualification. The explorers and travellers from this period were Westerners and their exploits were determined by and an integral part of colonial expansion (Polezzi, 2006). Starting with Napoleon's campaign in Egypt (1798–1801) and Lewis and Clark's transcontinental expedition (1804–1806), the century was characterised by an intensification of exploration. Either officially backed, or sanctioned by quasi-government bodies like the Royal Geographical Society (established in 1830), the goal of these explorers was territory, whether it be in the Americas, Asia, Australia or Africa. In a notable break with the traditions of previous Western discoverers, the shift in emphasis in the 19th century was from sea voyages towards walking across the interiors of these continents.

Matching this change in the form of exploration was a fundamental shift towards the romanticisation of the explorer ideal. In 1867, the African explorer Winwood Reade noted with much satisfaction that, 'when a great traveller enters a London drawing-room there are more rustling of flowers, and whispering behind fans, than welcome the novelist or even the poet' (quoted in Driver, 2001: 90). This popularisation of the heroic explorer was a function of rapid urbanisation in both Europe and the USA. For city dwellers following sedentary lifestyles, there was an unquenchable thirst for the vicarious pleasures of adventure and daring on the far-away frontiers. Ironically, 19th century explorers focused on walking long distances just at the time when most urban people were shifting towards mechanised transportation (Ingold, 2004).

This passionate interest translated into a massive increase in demand for adventure books. These included both the travel narratives of real explorers and a vast library of fictional accounts, ranging from Jules Verne and Henry Rider Haggard to the hack writers of dime novels (Laing & Frost, 2012). Apart from an extensive array of new titles, this was a period when hitherto obscure historical accounts were republished and appreciated. Thus, for example, the writings of Ibn Battuta, the 14th century Muslim traveller through the Mediterranean and Asia, were translated into French in 1853 and then into English in 1929 (Dunn, 1986).

Reading about heroic explorers reaffirmed the values of colonialism and was accordingly encouraged by schools and teachers (Parkes, 2009). As Robinson notes, 'the British Empire provided novelists with new dimensions of fiction ... bringing heroic adventure and exotic cultural encounters back to an increasingly literate Britain' (Robinson, 2002: 60). Many of these writers had been colonial officials (for example, Henry Rider Haggard), soldiers

(P.C. Wren) or journalists (Rudyard Kipling). They were conscious of a strong demand for stories of expanding and exploring the empire and that many of their young readers would potentially emulate their literary heroes by journeying to the colonies.

By the 20th century, much of the world seemed explored. The missing blanks in maps of Africa, Asia, South America and Australia had been filled in. The decline of empires reduced opportunities and swept away the culture that had valorised exploration as conquest akin to imperial military victories. In the early 20th century, there was a shift towards explorers filming and photographing their discoveries, particularly traditional societies. This allowed a much more vivid representation of their experiences for consumption by Western audiences (Gordon *et al.*, 2013). Yet, despite these changes, the myth of the explorer has remained, albeit modified. Western prosperity and cheap airfares have opened up much of the world for explorer travellers. Less inclined to be part of official scientific expeditions, many are influenced by the mystique and romance of discovery and adventure. The modern model is now more like a Patrick Leigh Fermor – not literally an explorer – but using travel to seek different experiences and even personal transformation.

The Explorer Traveller Discourse

This work is exploratory, in that there has not previously been any full-scale examination of the explorer traveller. Instead, the tourism literature touches upon it incidentally from time to time, often pointing to it as an area for further research. Rather than any fully developed theoretical framework, the value of this literature is in identifying key themes, issues and questions, providing not so much a roadmap, but rather general directions to guide us.

For us, the starting point is with the work of Erik Cohen (brought together in a 2004 collection of essays). He argued for a particular way of looking at the modern tourist, noting that they are driven to travel because they are:

> Interested in things, sights, customs, and cultures different from [their] own, precisely *because* they are different. Gradually, a new value has evolved [for the tourist]: the appreciation of the experience of strangeness and novelty. (Cohen, 2004: 38)

However, this driving force is often limited:

> Many of today's tourists are able to enjoy the experience of change and novelty only from the strong base of familiarity, which enables them to

feel secure enough to enjoy the strangeness of what they experience ...
Often the modern tourist is not so much abandoning his accustomed
environment for a new one as he is being transposed to foreign soil in an
'environmental bubble' of his native culture ... The experience of tourism
combines, then, a degree of novelty with a degree of familiarity, the security of old habits with the excitement of change. (Cohen, 2004: 38)

Cohen argued that the tourist experience is comprised of a continuum
of combinations of novelty and familiarity and proposed that this can be
understood through a typology of four tourist roles. These are: the organised mass tourist, the individual mass tourist, the explorer and the drifter.
His explorer:

Tries to get off the beaten track as much as possible, but he nevertheless looks for comfortable accommodations and reliable means of
transportation. He tries to associate with the people he visits and to
speak their language. The explorer dares to leave his 'environmental
bubble' much more than the previous two types, but he is still careful
to be able to step back into it when the going becomes too rough.
(Cohen, 2004: 39)

At the far end of the continuum, the drifter:

Ventures furthest away from the beaten track ... He shuns any kind of
connection with the tourist establishment, and considers the ordinary
tourist experience phony ... He tries to live the way the people he visits
live, and to share their shelter, foods, and habits, keeping only the most
essential of his old customs. (Cohen, 2004: 39)

Originally the drifter was a limited phenomenon. However, in recent decades
it has grown significantly as an appealing style of living and travel for
Westerners. Best represented as backpackers, Cohen described them as *mass
drifters* and *nomads from affluence*. He argued that the modern version is 'not
really motivated to seek adventure and mix with the people he visits. Rather,
he often prefers to be left alone to "do his own thing" or focus his attention
on the counter-culture, represented by the other drifters whom he encounters' (Cohen, 2004: 58).

Cohen's arguments are echoed among other researchers trying to
explain the movement away from mass tourism. Lewis and Bridger (2000)
argued that postmodern consumers are becoming more individualistic and
accordingly looking for novel experiences to express and satisfy that

tendency. Urry (2002) linked the demand for new and extraordinary experiences to a Western culture which demands gratification of desires, often impatiently. Similarly, Feifer (1985) saw a reaction to the boredom and sameness of mass tourism, resulting in savvy independent travellers wanting to have very different experiences, plus now having the time and resources to achieve this.

Other researchers looked back to the influence of the Romantic tradition and how that translated into the modern world. For Romanticists, wild places such as mountains, deserts, seascapes and forests were seen as providing spiritually enriching experiences. The antithesis of the crowded cities, they allowed for seclusion and solitary contemplation. The more spectacular that nature was and the more it was untouched by human hands, the greater the potential was for the intrepid visitor to experience the sublime. Such a view linked wild nature with transformation and self-exploration.

Exclusiveness was an important ingredient. This was not travel for the masses. Too many people visiting the same place made the sublime unobtainable. Instead, this was constructed as an elite experience. One needed the education and wealth to fully appreciate the benefits; indeed, being able to embrace the sublime in nature was a marker of status and good taste (Haynes, 1998; Sage, 2009; Urry, 2002).

A number of researchers emphasised that modern travellers were primarily motivated by a quest for authenticity, both in relation to self and setting (Cohen, 2004; Laing & Crouch, 2011; MacCannell, 1976; Selwyn, 1996; Wang, 1999). This was seen as fuelling a desire to go further and further than others, to push physical and inner boundaries. As Cohen argued, there was now 'an intensified quest for authenticity and otherness – whether in the remaining traces of the past, in the challenges posed by the extreme, often inhospitable margins of the Earth, and, especially, beyond the Earth, in space' (Cohen, 2004: 321). This desire was now matched by a greater availability of resources, as technological development and economic prosperity were providing greater opportunities for tourists to go to destinations and have experiences that had hitherto only been available to explorers, scientists and astronauts (Crouch et al., 2009).

Two qualitative studies that examined inner development through adventure-style commercial expeditions are worth noting. Gyimóthy and Mykletun (2004: 856) found that participants on Arctic snow-scooter and skiing treks were seeking 'intense physical and psychological challenges ... [which] can trigger deep euphoric experiences'. Furthermore, a key part of the psychological arousal was wholehearted engagement in play, make-believe and role-enactment to create a fantasy adventurer/ explorer experience. Arnould and Price (1993) interviewed participants in rafting trips on the Colorado River, identifying what they called 'extraordinary experiences':

> It is recalled easily for years after, but, because of its considerable emotional content, it is difficult to describe. People sometime report that it changed them forever. It is magical. As such, satisfaction with river rafting, a hedonic encounter between customer, guide, and 'nature,' does not seem to be embodied in attributes of the experience such as amount of time spent freezing in wet clothes, uncomfortable toilet facilities, bad food ... Rather, satisfaction is embodied in the success of the narrative, an interactive gestalt orchestrated by the guide over several days' journey into the unknown. (Arnould & Price, 1993: 25)

This quest for authenticity was illustrated by some travellers who sought to recreate other's journeys and experiences (Laing & Crouch, 2011; Laing & Frost, 2012). These recreations served two roles. The first was a commemoration honouring those past heroes, in some cases even justifying them and seeking to rewrite how they are seen in the history books. The second was that they allowed the re-enactors to also undergo a transformation, testing themselves against adversity and encouraging self-reflection, even escape.

Recent exploratory work by Crouch (2013) applied concepts from evolution and sociobiology to tourism studies. His argument was that the need for travel was important among prehistoric peoples, providing benefits of new resources, trade and status versus some risks. Those who were adventurous were genetically successful and this may have been passed on to their ancestors. Reflecting that there are some people who seem highly restless and have a high propensity to seek adventurous travel, Crouch pondered whether this might be part of their genetic make-up.

The role of the media in projecting an image of explorer travel as glamorous, worthy and desirable has barely been touched upon. Young (2009) argued for a *Circle of Representation*, where tourists seek to imitate what is produced in the media. This might entail, for example, reproducing photos of scenes and experiences as they are presented in guidebooks. Taking this concept further, it is argued that explorer travellers are fascinated in following, if only in part, the travel they read about in fiction and other travellers' narratives – whether that be in books, or increasingly via blogs and other social media (Dann, 1999; Howard, 2012; Zurick, 1995). This is an example of *intertextuality*, an 'interrelationship [between texts] through either hidden or open references' (Månsson, 2011: 1637), particularly as applied to tourism. The referencing of previous sources, whether directly quoted or alluded to, or merely echoed (known as *transtextuality*; see Sankaran, 2008), opens up questions about the objectivity of travel narratives, including the blurring of fact and fiction (Norman, 2009). As Polezzi observed: 'just as travel crosses boundaries, cultures and languages, so travel writing produces texts which are marked by alterity, by distance, and by multiple allegiances, crossing fact and fiction, autobiography and description, ordinary life and extraordinary adventure' (Polezzi, 2001: 1).

Exploratory research by Richards and Wilson (2004) surveyed a small number of backpackers, finding that books were more influential than film. This study identified key authors – 'backpacking icons' – including Ernest Hemingway, Jack Kerouac, Hunter S. Thompson, Bruce Chatwin, Paul Theroux, Michael Palin and Bill Bryson (Richards & Wilson, 2004: 47). The reading of books not only inspires the desire to visit particular destinations, but fixes common notions that travel may be transformative, a necessary rite of passage and even excitingly dangerous (Frost & Laing, 2012; Laing & Frost, 2012). There was also a growing focus on how media portrayal of travel to remote places often highlights this concept that the visitor is transformed by the experience, a common theme noted in fictional films set in the Australian Outback (Frost, 2010).

Cronin's (2000) work on *translation* considered the intersection between travel and language, particularly the way in which encounters with *the Other* are often mediated by a translator – who could be a guide, a companion or a writer – 'who straddles the borderline between the cultures' (Cronin, 2000: 2). Writers of travel narratives therefore play a role in translation, interpreting what they see for their readers, and thus 'contribute powerfully to the construction of national cultures through language' (Cronin, 2000: 22). It is noted that these accounts are often underpinned by political agendas and are therefore a product of their own time. Upon coming home, Cronin argued that the traveller's narrative returns 'in memory and prose to the places and experiences of travel' (Cronin, 2000: 35), albeit in their own language and from their own cultural standpoint.

Some researchers focused on the potential negativities of explorer travel. In what we might now term an autoethnographic approach, geographer Zurick analysed what he terms adventure travel. This involves young Western tourists seeking to immerse themselves in the cultures and wilderness areas of developing countries. An intertextual range of historical and modern media inspire them:

> Historical accounts of early explorers . . . now guide the modern travelers – as pseudo maps into the exotic world and as references against which contemporary experiences can be measured. Old and new stories shared

among travelers, narratives of the journey, establish complex circuitries of adventure travel. (Zurick, 1995: 54)

Building on the work of Cohen, Zurick went further in questioning the mores and impacts of this phenomenon. In cataloguing a range of influences on traditional societies, he argued that:

> Ironically, adventure tourism carries with it the very defilements that such tourists wish to escape ... [for they are] unwitting bearers of the worlds they have temporarily abandoned ... what were once social exchanges are now economic transactions, sacred rituals change to secular ceremonies, native dress becomes curio costumes ... adventure travel may not produce the types of economic and social shifts that characterize full-blown mass tourism, but its impacts ... nonetheless can be considerable. (Zurick, 1995: 2–3)

Zurick is an appropriate writer with which to close this discussion of the explorer travel discourse. Intensely personal, he not only pondered how this form of tourism changes the world, he judged it. He was not alone in this; he simply stood out in not disguising his views. This was well illustrated in one of his accounts of trekking in the Himalayas. At night he reached a simple local lodge. He was joined by three similarly exhausted Australians. Sitting around a stove, 'after a few pleasantries and obligatory comments about steep trails but lovely views, everyone grew silent and gazed into the flames'. Zurick felt a common bond, 'such complete anonymity and the severance of ties with everyday life are the seductions to which travelers succumb' (Zurick, 1995: 12–13). After a while the Australians went to their room. Within a few moments, Zurick's peace was shattered, for they started loudly playing a recording of an Australian rock group (AC/DC? Angels? Midnight Oil?). Zurick was appalled. This was not the right way to behave! Furthermore, he noted, 'the innkeeper recognized the music and knew the band's name; I did not' (Zurick, 1995: 13). Such subjective judging of the motivations and behaviours of tourists is a common theme throughout this literature.

A Typology of Explorer Travellers

There are a wide range of travel experiences which can come under the broad heading of explorer travellers. To better understand this phenomenon, we present a rough typology. This is intended to be only general and descriptive, rather than analytical and prescriptive. In presenting this, we are simply trying to convey the range of travel experiences that might come under the heading of explorer travel. Furthermore, it is clear that in any typology there will be a great deal of overlap between the various types. With those qualifications in mind, we suggest the following nine categories, providing examples of each.

Official expeditions

Sanctioned and funded by governments or scientific organisations, these have the gravitas of officialdom. The goals are the exploration of new lands (often with a view to colonisation) and scientific discoveries. Their nature and status means they are often in the command of a military or naval officer. Examples include: Roald Amundsen, Robert Scott, Ernest Shackleton and Douglas Mawson in the Antarctic; Speke and Grant searching for the source of the Nile; Lewis and Clark crossing America; the 1870 Washburn Expedition to Yellowstone; and the tragic failure of Burke and Wills in the Australian Outback. The exploration of space (so far) has followed this pattern, for example with the Apollo missions to the Moon. However, this now seems to be changing, with an increasing privatisation of space exploration.

Unofficial discovery expeditions

Similar in nature and goals to official expeditions, these lack government involvement. Instead, funding and impetus may come from the media, private businesses and public donations. Examples include Henry Morton Stanley's expedition to find Livingstone, which was funded by the *New York Herald*, Thor Heyerdahl and the *Kon-Tiki* expedition, and many current-day explorations funded by *National Geographic*. In such circumstances, there is often a strong emphasis on a media production, such as a book, television documentary or online blog.

Conquerors

The mission is to be the first to conquer some sort of natural barrier, to break a record and potentially be listed in the *Guinness Book of Records*. While this mission may be dressed up as scientific, the core goal is to be first – to go where no one has been before. Examples include Edmund Hillary and Sherpa Tenzing Norgay being the first to climb Mt Everest and Felix Baumgartner, who in the Red Bull Stratos in 2012 set the records for the highest altitude for a manned balloon, a parachute jump from the highest altitude and the greatest freefall velocity.

The personal quest

The journey accomplishes some personal goal. The emphasis is on the individuals testing themselves. In some cases there may be a spiritual dimension, making this akin to a personal pilgrimage. The focus on the personal shifts the expedition goal away from objective concepts of discovery or conquest. Instead, the individual is breaking through personal boundaries which they have identified for themselves. The goal may have already been achieved by others; indeed they may form the inspiration that this person can also duplicate this. Examples include the continuing fascination with climbing Mt Everest, walking the Kokoda Track or the Camino Way, or visiting a particularly dangerous region.

Change of life

Related to the personal quest, the difference is that the quest has been sparked by some life-changing experience. An illness or bereavement may be the catalyst, with the individual developing a journey as a targeted means of assisting in the recovery. An example is Warren Macdonald (1999), a climber who overcame the loss of his legs (see Macdonald, 1999), and the blind mountaineer Eric Weihenmayer (2001).

The re-enactment

A re-enactment of a historical explorer's journey is undertaken. The past exploration may have fascinated for years, leading to a desire to emulate the achievements of a hero. Variations may be the honouring of the past explorer at some auspicious date, such as a commemorative anniversary, or the desire to test some contentious aspect of the previous expedition. Examples include: Heyerdahl attempting to prove that South Americans could have reached Easter Island; the recreation of Scott's ill-fated expedition to the South Pole by Roger Mear and Robert Swan; and retracings of the similarly disastrous Burke and Wills expedition to cross Australia. The latter was recreated by Tom Bergin in 1981 and again by the Royal Society of Victoria to commemorate the 150th anniversary in 2010.

Quirky or comedic

There is some novel aspect of the expedition, ranging from the whimsical to the outright absurd. Such an affectation demonstrates that the traveller is different in having a 'devil may care' attitude, possibly even poking fun at the seriousness of other travellers. Examples include travels by comedians (which of course are produced as television series) and the use of bizarre modes of transport, such as unicycles, skateboards, and so on. A specific example was the *Journey to the Centre of the Earth* by cousins Richard and Nicholas Crane. Rather than attempting to duplicate the novel by Jules Verne, their aim was to reach by bicycle that point in Central Asia that was furthest away from the sea in all directions.

The independent traveller

Probably the largest grouping in terms of numbers, this is akin to Cohen's concept of the explorer. Essentially tourists, they are travelling independently, developing their own itineraries. They are exploring in that they are finding their own way and constructing their own experiences. A subcategory is that of the *flaneur*, who is a wanderer with no set plan. Both concepts may be blurred, as in a traveller following a set itinerary of places to visit, but once there, wandering without much structure. Examples of this independent type are backpackers and gap year travellers.

The commercial expedition

The journey is staged by a commercial operator, with travellers paying to take part. While essentially a guided tour, the participants may not wish to see it in that way. Indeed, there is often added risk due to the uncertain nature of relationships between the guides and guided. Examples include commercial climbs of Mt Everest and other peaks and expedition tours to Antarctica.

Our typology follows a fluid continuum of seriousness of purpose. At one end are professional explorers, sometimes officially sanctioned. Their expeditions are expensive and have clear externally directed goals, either scientific or political. At this end, participants are professionals, seeing this as a career (with a clear progression) and engaging in years of training. Further along the continuum comes a semi-professionalism. The goals are broader; the level of support varies. Many see this as a temporary engagement, a journey that will be arduous and require much preparation, but which is something they will only do for a while. In many respects this equates with the concept of *serious leisure* (Stebbins, 1992). Some in this middle ground have aspirations of moving higher; their expedition is a stepping stone for gaining fame or media attention and becoming a professional. In this middle part of the continuum, the goals have tended to become more internally directed, focused on what the individuals wish to achieve for themselves. At the far end of the continuum comes a far larger number of people who are incorporating elements of exploration and adventure into their experience, but who are funding this themselves and are essentially tourists. They may, of course, do this very regularly and we may view this as a form of serious leisure. For these groupings, the goals are entirely internal.

Some Themes

Our brief overview of the explorer traveller discourse and our rudimentary typology highlight some themes which we aim to explore further in this book. These include:

(1) There is an *Explorer Myth* that strongly influences modern independent tourism. Based on historic accounts and contemporary narratives, it delivers a promise of adventurous and testing travel. The explorer is heroic and on a journey comparable to that of mythic heroes as outlined by Campbell (1949). Many modern travellers seek to duplicate this myth, undertaking their own explorations modelled on the exploits of others.

(2) Cohen's explorer and drifter types are dynamic, changing over time. The drifter, for example, has transformed from a small subculture into a massive phenomenon. New areas of travel are constantly opening up, for example into space and deep underwater.

(3) There is a tendency to the extreme in travel. Where dangerous activities and regions were once the province of highly trained specialists and professionals, more and more it is ordinary people who want to have these experiences. Many report that they are going to the extreme in order to push their own personal boundaries. Their quest is for inner benefits, ideally a major transformation.

(4) If exploration travel is transformative, how does that work? What are the mechanisms of transformation? Are they physical – exposure to discomfort and danger, or even the rhythms of focusing on a repetitive task such as walking?

(5) Explorer travellers are subject to a wide range of influences, particularly as to destinations, experiences and the possible effects upon them. Of particular importance are the intertextuality of works of fiction (film, television, novels) and narratives of past explorers and explorer travellers. Many of these conform to the structure of *The Hero's Journey* as developed by Campbell (1949). This concept posits that around the world, mythological heroes followed common patterns of engaging upon a testing and transformative journey. Such a structure still seems

relevant to understanding the behaviours of modern-day explorer travellers.

(6) As the numbers of explorer travellers grow and their reach extends more and more into remote areas, they are having a greater impact on host communities and cultures. This raises issues of sustainability, globalisation and the negative impacts of tourism.

(7) The discourse on explorer travellers – like tourism in general – is subject to cultural biases. There is a dominant Western paradigm. Indeed, explorers are often viewed in terms of the Anglophone world, although recently there have been arguments that this requires a broader view. There is a strong danger that other voices are not heard, or are mediated through the filter of the dominant culture (Forsdick, 2005; Polezzi, 2001). Similarly, there are issues of gender. Is explorer travel essentially masculine? There are women explorer travellers, but how do we interpret them? Do we need quite new and different approaches to researching them?

(8) The role of *translation* in the explorer travel experience is complex and nuanced. Encounters with the Other and misunderstanding of language and culture can be 'disorienting and threatening' (Cronin, 2000: 3). How do explorer travellers deal with or resolve these differences through their narratives? And does this discourse of cultural divergence broaden the traveller's (and their audience's) horizons or merely reinforce stereotypes and power imbalances?

(9) Drawing on research in human genetics and prehistoric anthropology, questions are being asked as to whether many of us are genetically programmed to wander restlessly. Are we subject to an *explorer gene* – first developed hundreds of thousands of years ago – that is still active in our modern world?

Scope and Structure of this Book

In examining the myth of the explorer and how that influences modern tourism, this work builds on our previous work *Books and Travel: Inspiration, Quests and Transformation* (Laing & Frost, 2012). In that text, we analysed how books – whether fiction or factual – reinforced the idea that travel was adventurous and exciting and possibly even transformative. In considering these issues we looked at a wide range of genres, such as historical fiction, crime, westerns and children's books. One of the most influential genres was that of the explorer's account. These could be fictional – as in the novels of Jules Verne – or based on real adventures. Of the latter, we examined 13 explorer narratives. In doing that, we realised that we had only scratched the surface, that there was much more that could be done with this particular genre. For us, it represented unfinished business and we felt that we needed to proceed to a full-length examination of how the concept of the explorer pervades tourism.

There were also connections with previous research. Jennifer had examined the motivations of frontier travellers – particularly how they were influenced by earlier explorers and travellers and were often fascinated by recreating their travels and travails (Laing & Crouch, 2011). As an active member of the Mars Society (www.marssociety.org), she is interested in the promotion of space travel and exploration. In 2002, she organised the Australian speaking tour of the Apollo astronaut Dr Harrison Schmitt, one of the last two men (and the only scientist) to walk on the Moon. In 2003, she spent two weeks living in the Mars Society's simulated (analogue) Mars base in Utah as a member of the expeditionary team (Figure 1.1).

Warwick's research on rainforests drew in many connections with historic explorers – particularly botanical expeditions and the tropes of Western discovery, conquest and exploitation. In his work on tourism in the Australian Outback, he had examined how cinema commonly focused on its spiritual (or even magical) powers in transforming the lives of those who ventured deeply into it (Frost, 2010). For us, these strands of research come together in examining how people imagine exploration and travel.

Figure 1.1 Analogue research for the Mars Society in Utah
Source: J. Laing.

This book is divided into four parts. The first (covering Chapters 2–5) is titled The Hero's Journey. In this we examine the Myth of the Explorer using Campbell's framework. He argued that many cultures around the world had common myths involving a hero's journey, that these followed a similar structure and that this structure was often applied in modern story-telling. We apply this framework to a wide range of explorer narratives – both in published books and in the interviews we conducted with modern-day explorer travellers. We argue that these explorers' journeys follow Campbell's structure and that explorers may be understood in these mythical and heroic terms. Furthermore, this elevation of explorers to a mythical status provides the motivation for many travellers to undertake their own heroic journeys.

Part 2 examines how exploration may be reimagined through fiction and re-enactment. Chapters 6 and 7 consider a range of fictional works, primarily novels, but also some cinema and television. In such fictional renderings, certain elements of the explorer are emphasised and some exaggerated. As Chapter 6 illustrates, there is a focus on heroism and adventure, but this is also balanced with tropes of foolhardiness, obsession and exploitation. In some instances, the explorer has even become a figure of fun. Chapter 7 concentrates on one fictional storyline – that of the stranded explorer. Here there is a strong fascination with the hero being tested by being prevented from returning. In some cases, their predicament is highly romanticised; in others these fictional works focus on the dangers of poor decision making and tensions within the group. Chapter 8 examines some instances where past explorations have been re-enacted. In some cases this is to recreate the dangers and adventures of iconic journeys, in some to honour those who have perished, and in others to attempt to resolve mysteries that still remain.

The third part extends our discussion of the explorer myth to trends in modern tourism. Chapter 9 looks at travellers who attempt to immerse themselves in exotic cultures, adopting their dress and customs and in some cases even marrying into them. Such attempts represent a desire to cross forbidden or difficult borders, to leave behind modernity and *cross over* into a more traditional world that has been constructed as more authentic and desirable. This is highly contested ground, with these immersive travellers suggestive of neo-colonialism as they appropriate other cultures for their own ends. Chapter 10 considers organised commercial tours, generally mar-keted as safaris and treks. These are examples of Pine and Gilmore's *Experience Economy*, the staged and scripted performance of experiences for profit. In examining these tours, we consider the paradox between the ide-alised independence of explorer travellers and the constraints of packaged expeditions.

The fourth and final part looks to the future. Chapter 11 examines how future trips to Mars might be conducted, particularly how commercial travel might one day be achieved. Chapter 12 concludes with a critique of the experience economy and argues that, for an increasing group of savvy modern-day travellers, such scripting and pre-packaging is deeply unappealing. Our argument is that it is exploration and discovery that travellers nowadays crave. We put forward the proposition that exploration travel – based heavily on explorer narratives and the promises of personal challenges and change – is a major trend in future tourism.

Part 1
The Hero's Journey

2   The Call to Adventure

Why individuals are moved to set out on quixotic and extraordinary journeys can be conceptualised under the broad heading of the *Call to Adventure*. Campbell labels this 'the signs of the vocation of the hero', where he or she accepts their destiny as an adventurer. It is the start of 'the first great stage' of the hero's journey – the *separation* or the *departure* (Campbell, 1949: 36). In this chapter, we consider some of the key motivations behind the call to adventure – what makes a person willing to spend vast sums and undergo high personal risk and danger in order to travel to remote places – before turning to consider the departure more fully in the following chapter. Unlike Campbell's hero, the travellers we discuss in this chapter don't generally refuse the call once they hear it, although it may take them years to actually embark on the journey.

The Serendipitous Call

Pinpointing the exact moment when a mind turns itself to adventure is not always possible. An invitation to embark on a journey or expedition can be serendipitous or casual. Eric Newby (1958) commences his book, *A Short Walk in the Hindu Kush*, with the convention of a telegram, which he sent to his friend Hugh Carless at the British Embassy in Rio. It reads, 'Can you travel Nuristan June?' These days it would be an email or a text. Elizabeth Best suggests that her pilgrimage along the Camino Way with Colin Bowles 'came about randomly in conversation ... over the phone' (Best & Bowles, 2007: 3), while Caroline Hamilton decided to go on a polar trek after meeting adventurer Pen Hadow by chance at a party (Hamilton, 2000).

For others, the decision to travel occurs long before they have the means to do so, often during childhood or early adolescence (Laing & Crouch, 2009a), and niggles at them until they can heed adventure's call. Catherine Hartley, a polar trekker, notes that she has 'always longed to travel, ever since I was a child. I wanted to go to places that were undiscovered. I spent hours pouring over my mother's atlas of the world' (Hartley, 2002: 11). Jonathan, a polar trekker we interviewed, gained 'the traits of exploration . . . by osmosis' through the books his parents read to him as a child. Robyn Davidson first visited India in 1978 and a visit to a camel fair implanted the idea of accompanying nomads on their journeys. However, it would not be until 1992 that she took the plunge and embarked on such a venture (Davidson, 1996). Aimé Tschiffely justified his horse ride from Buenos Aires to Washington, DC with:

> For a long time I had felt that a schoolmaster's life, pleasant though it is in many ways, does not afford much prospect and is apt to lead one into a groove. I wanted variety: I was young and fit: the idea of this journey had been in my head for years, and finally I determined to make the attempt. (Tschiffely, 1932: xvi)

John Muir also experienced the call to adventure at an early age. He writes:

> Boys are fond of the books of travellers, and I remember that one day, after I had been reading Mungo Park's travels in Africa, mother said, 'Weel, John, maybe you will travel like Park and Humboldt some day'. Father overheard her and cried out in solemn deprecation, 'Oh, Anne! dinna put sic notions in the laddie's heed'. (Muir, 1913: 103)

Campbell notes that the hero's journey may commence with a simple mix-up or error, which catapults the hero into meeting their destiny, described as a *blunder*. He gives the example of the princess who loses her ball while playing in the garden and finds a frog in *The Frog King*. In contrast, the adventure traveller does not blunder into a journey, or at least does not admit to it in their narratives. While it appears to be acceptable, even admirable, to refer to the genesis of some travel as unexpected or the result of *chance*, it might be beyond the pale for explorers to acknowledge that their journey was accidental. This would affect the prestige or status of the traveller, which is such a strong motivation (Laing & Crouch, 2005, 2011).

Objects or Talismans

An object may inspire a journey and become a type of *talisman* or symbol of good luck in the eyes of the explorer. In mythology, this often has magical properties. For example, Theseus was given a magic ball of thread by Ariadne, to help him escape from the Labyrinth after slaying the

Minotaur. The British explorer Percy Fawcett was given a wooden idol by his friend Henry Rider Haggard, author of *King Solomon's Mines* and *She*. Fawcett regarded it as a lost relic of the City of Z, the mythical city he believed existed in the Amazon jungle, despite the fact that a number of experts had labelled the idol a fake (Grann, 2009). In 1925, on an expedition during which his whole party disappeared, he was asked by a ranch owner about his reasons for travelling to the Amazon. In reply, he showed the man his idol.

Some modern adventurers carry an object owned by an earlier explorer. This may take on a sacred quality for the traveller, given that there is normally only room for a few personal possessions amid the necessities required for these arduous journeys (Belk, 1992) and some are forced to reduce the weight of their kit in order to be able to keep going (Best & Bowles, 2007). Henry Worsley (2011) took Shackleton's compass with him on a re-enactment of one of his South Pole journeys and often touched it, as a source of comfort and a link to his hero.

Childhood Influences

The call to adventure is often nurtured from childhood or adolescence, forming a dream or fantasy that might only be able to be realised when the individual becomes an adult (Laing & Crouch, 2009a). Some of this early interest in adventure can be linked in particular to the influence of parents and teachers (Celsi *et al.*, 1993; Laing, 2006). Families may pave the way for their children's adventurous travel by encouraging outdoor pursuits. A number of the people we interviewed joined Scouts or Outward Bound courses, took part in the Duke of Edinburgh Award Scheme or participated in school camps, which they felt moulded them, developed character and taught them the skills which they built upon as an adult (Klint, 1999). According to Helen, one of our interviewees, these types of activities 'sparked an idea that there were lots more things possible beyond my sheltered little world'.

Mountaineer Conrad Anker was taken into the outdoors by his family from a young age and learned how to function independently in the wild: 'Nowadays a lot of people come to the sport by training in a climbing gym ... They may know how to pull up an overhang, but they don't know what an afternoon cloudburst can do to you if you don't pitch a tarp. I learned that at a ripe young age' (Anker & Roberts, 1999: 97). David, a mountaineer, spoke enthusiastically to us about his school, which was 'quite pro-active, very enthusiastic and encouraging to us to get out and do a lot of different sports ... We had a very good teacher in charge of adventure sports ... At fifteen, to go off into the snow and live in the snow for a week'. Ann Bancroft's teacher Pat became a mentor and encouraged her sporting and academic abilities, such that she wanted to make 'a difference in kids' lives the way Pat had made a difference for me' (Arneson & Bancroft, 2003: 111).

Mountaineer Peter Hillary was encouraged by his mother, rather than his more celebrated father, 'to get out there, into the adventuring game'. Hillary notes: 'It was Mum who organized my first alpine climbing trip, to Mount Ruapehu, with the Alpine Club in the late sixties, when I was maybe fifteen ... Mum saw I needed to go out on a limb; that I needed to challenge and express myself as a way of breaking out of my excruciatingly self conscious and awkward shell' (Hillary & Elder, 2003: 223). Perhaps she also encouraged Hillary in these pursuits as a way of allowing him to break free of the burden of being a famous adventurer's son. Hillary explains how his father's bloodline flows through his veins:

> [H]ere we are Hillary and son. Just as there are family lines of doctors, lawyers, farmers, we became a small mountaineering concern. The first father and son to climb to the top of the world, actually. I am following in his footsteps, in many ways, and in many ways deeper and more important than our connection to the hills or the thrills. They are but the expressions of breeding and blood.... (Hillary & Elder, 2003: 210–211)

Michael told us of his father's influence on his polar and desert trekking exploits: 'I certainly think that my father's difficulty in communicating that he'd felt I'd done well has driven me to try and do more things.' He also explained how he grew up in Asia in a more freewheeling and exotic society and saw travel as his gateway to recapturing this sense of freedom and escape that he knew as a child:

> I grew up in the Far East as a kid and I saw some quite different things and [then] I found myself back in a fairly regimented world that was British public school and all that kind of stuff, having grown up for eleven years in Malaysia and Singapore. I just felt there was a dimension to life that was missing when I went back to that world from the one I'd had. And I wanted to get back to living life as more of a series of experiences than such a regimented structure, which is British life to a certain extent.

Childhood games may be a form of role-play with respect to future adventure travel and exploration. Zurick (1995: 61) notes that fantasy 'predates

the journey and continually propels it along new courses'. Geoff, for example, used to pack his Action Men figures to take on every car journey and
labels this 'my earliest days of expeditions'. Another adventurer, Graham,
told us: 'I grew up in the country, and I'd go on all-day horseback adventures by myself pretending I was an explorer. Typical things that a young
fellow does out in the country. So I've been doing this type of thing from
an early age in a psychological sense. Going exploring.' Similarly, mountaineer Jon Muir writes: 'I spent all my spare time out in the bush. Exploring,
adventuring and making spears, bows and arrows, claypots and cubby-
houses. The bush was where I was happiest and where I felt I belonged'
(Muir, 2003: 4).

Childhood influences about travel can be long lasting. Ranulph Fiennes
writes that his partner on his South Pole unsupported expedition, Mike
Stroud, 'equated his adventures with a more intense version of the pleasures
he found as a boy from mountaineering, hill-walking and rock-climbing'
(Fiennes, 1993: 27). Journalist Jon Krakauer observed, after receiving an invitation to join an expedition to climb Mt Everest, 'Boyhood dreams die hard,
I discovered, and good sense be damned' (Krakauer, 1997: 28).

However, the biggest single childhood influence is that of the media,
particularly books (Laing & Crouch, 2009a; Laing & Frost, 2012). As Alex
observes: 'Reading's a big [influence] because that kind of fuels the dreaming
and it gives you more pictures to dwell on.' John Goddard's parents encouraged his love of learning and natural curiosity as a young man through the
gift of a set of the *Encyclopaedia Britannica*:

> Reading extensively in each of the twenty-four volumes opened up the
> world for me as never before – an inexhaustible treasure house of fasci-
> nating information, covering every conceivable topic, that stimulated my
> interest in countless other subjects. (Goddard, 2001: 242)

Such a view runs counter to Butler's (1990) contention that, as reading
declines as a popular pursuit, visual forms such as cinema will have a greater
effect in shaping destination choice and travel behaviour. It is important to
understand that books have not lost their power as a siren call to travel and
they still shape the way we *understand* travel, as well as *how* we travel and
our expectations of how travel will *transform* us (Laing & Frost, 2012). They
are *organic* agents of images about travel, which are not linked to particular
destinations or tourism campaigns, but arise and are shaped through our
personal experiences and backgrounds.

Literature resonates for adults as well as children. For example, kayaker
Scott Lindgren, according to his team-mate Heller, 'stole a copy of Kingdon

Ward's *Riddle of the Tsangpo Gorges* from the Auburn library and devoured it. The book told a tale that made Scott almost sick with desire' (Heller, 2004: 89). Patrick Woodhead, with reference to the Australian polar explorer and geologist Mawson, explains the inspiration he derives from reading about his heroes: 'Characters like that are just so appealing. Not only to school-boys searching for heroes, but to anyone who pushes themselves farther than they previously felt possible' (Woodhead, 2003: 309–310).

This is not to say that films are not also influential on explorer travellers, but not apparently to the same degree as books. Cinema or television can be characterised as *pull factors* for tourism (Riley & Van Doren, 1992) and may bring adventurous journeys to vivid life (Gordon *et al.*, 2013). As Morkham and Staiff (2002: 300) observe: 'The ability for film to transport audio-visu-ally other worlds (other places, other times) into the present is unique ... bringing worlds to the spectator that may otherwise remain out of reach.' Alex, for example, sought out *paradise* during his travels, exemplified for him by the location for the movie *The Beach*. Zurick (1995) notes how paradise or utopia is generally sought in faraway places rather than close to home, because of its hallmarks of mystery and elusiveness. Ironically, the physical setting for *The Beach* is now a recognised tourist haunt and an example of paradise that has been spoilt by mass tourism and over-commercialisation (Law *et al.*, 2007).

Television was also mentioned by our interviewees. Simon, a diver, suggests:

> I think people of my vintage often say that the thing that turned them on to diving was the Jacques Cousteau specials, the Lloyd Bridges *Sea Hunt* stories. It was terribly exciting and adventurous and seemed a little more accessible in some ways than some adventurous pursuits like climb-ing Mount Everest because it became more immediate through the medium of television.

Similarly, Jack recounted that:

> When I was a young boy, I saw documentaries and I saw adventurers who went into the jungle in South America and had first contact with humans there, with Indians...and so I was very interested, very thrilled about expeditions, especially into the jungle, with travel [involving] people.

The power of photographs is also strong and can leave an impact which is still felt many years later. Several climbers in particular mention the motivation provided by photographs or pictures. Bear Grylls, who climbed Everest in his early twenties, says that:

> I was eight years old when my father gave me a huge and wonderful picture of Mount Everest. From that moment onwards I was captivated. I would sit there trying to work out the scale of the huge ice fields I saw in the foreground, and to judge how steep those summit slopes would really be. My mind would begin to wander, and soon I would actually be on those slopes – feeling the wind whip across my face. From those times, the dream was being born within me. (Grylls, 2000: 4)

Robert Swan, who recreated Scott's fatal expedition in 1912 to reach the South Pole, similarly describes the inspiration he derived from a photo of Scott he found in a library book at Durham University and how it piqued his curiosity about the ill-fated journey:

> My eye was caught by the cover of one particular book – *Scott's Men* by David Thomson. I stared for a long time at the picture of four men on the cover, their eyes sunk in their sockets, skin stretched tight over cheekbones; but it was less these physical manifestations of starvation and privation that told me they had suffered than the haunted expression in their eyes, a look that Thomson described as the 'flawless gaze of young men in the South.' What the hell had gone on, I asked aloud, to make that happen? (Mear & Swan, 1987: 16)

Those photographs of Scott were to haunt Swan and partly inspired his recreated journey:

> I wanted so much to bring all the photographs alive. I wanted to defend Scott's judgment and the noble sacrifice of Captain Oates, and in so doing, to uphold the tradition of Polar exploration and draw attention to the pressing need for conservation in the last great wilderness on earth. It was a tall order. (Mear & Swan, 1987: 18)

It has been said that to pore over a map *is* to travel (Marin, 1993; Morkham & Staiff, 2002). Ben Kozel, who journeyed the length of the Amazon River by raft, read atlases voraciously from an early age, and remembers 'always pausing longest on the maps of South America' (Kozel, 2002: xxiii). He also looked through maps and articles contained in his father's collection of *National Geographic* magazines, and comments about the

Amazon: 'I remember thinking how impenetrable that landscape looked, how it seemed to harbour a thousand mysteries, none of which would ever be solved' (Kozel, 2002: 242–243). Kozel's observation shows how mental maps can be created, linking texts as well as geographic pathways – 'imaginary routes constructed of oral and written narratives, not merely drawn by the cartographer's pen' (Zurick, 1995: 46).

Motivations Behind Exploration

The reasons for embarking on adventurous travel may be multifarious, largely unconscious and complex to unravel (Laing, 2006). Common tropes in explorer narratives are the challenge, overcoming obstacles, achieving goals, escape, self-discovery and honouring others.

Challenge

One important motivation is the desire to challenge oneself, both physically and mentally. A number of studies (e.g. Arnould & Price, 1993; Celsi *et al.*, 1993) refer to the element of challenge required in adventure, which is sought by individuals to 'test their abilities and achieve high arousal' (Gyimóthy & Mykletun, 2004: 873). It thus has a link with achievement (discussed below), as well as *flow* experiences, which refer to 'the complete involvement of the actor with [his or her] activity' (Csikszentmihalyi, 1975: 36), where the skills possessed by an individual and the challenges posed by the activity are finely balanced, leading to intense feelings of enjoyment and pleasure (Csikszentmihalyi, 1975).

Percy Fawcett refers to his 1925 expedition to find the Lost City of Z as leading the men to 'suffer every form of exposure . . . We will have to achieve a nervous and mental resistance, as well as physical, as men under these conditions are often broken by their minds succumbing before their bodies' (quoted in Grann, 2009: 11). Without hardship, the journey has no meaning and no purpose. In a modern context, Chris Bonington attributes his love of climbing to a combination of the seductiveness of risky activities, and encountering the unfamiliar and visiting remote places:

> I caught a bus and set out on my first mountaineering expedition . . . This was adventure. Tentatively, I was stepping into the unknown, had an awareness of danger – admittedly more imagined than real – and a love of the wild emptiness of the hills around me. (Bonington, 2000: 6)

Doug, also a mountaineer, told us:

> The challenge is the biggest part of it. I want to see if I can do it and if I
> think the easy route on a mountain is not going to pose a serious chal-
> lenge or worthwhile challenge, then I won't do that route ... Otherwise
> I'm just paying a lot of money and putting myself into a lot of danger
> for no real achievement at the end of the day. No gain.

Some of this challenge is so punishing as to be virtually masochistic.
Ranulph Fiennes and his partner Mike Stroud took part in a trek across
Antarctica in 1992/1993, which was so harrowing that they provided data
for medical research which could not be obtained any other way. This was
partly due to the ethical issues of obtaining informed consent, as well as
the difficulty of getting the average person to volunteer for such an assign-
ment. The descriptions of the suffering Fiennes and Stroud underwent
during their journey, such as the dramatic weight loss and physical injuries
suffered, are so extreme that they could be referring to victims of torture
or war when relating the deleterious effects on their bodies of their extreme
journey. Stroud says of Fiennes' legs: 'You can't pull anything with those.
They're Belsen-like' (Fiennes, 1993: 163), while Fiennes notes: 'With no fat
layer left, [Stroud's] hipbones were sticking into his skin' (Fiennes, 1993:
219). This level of suffering takes the pair into a realm that almost beggars
belief, given that their suffering was voluntary, and makes for disturbing
reading. Incredibly, Fiennes went back to the South Pole in 2013, with a
team of five, hoping to become the first person to cross Antarctica in
winter. He labels his latest adventure, *The Coldest Journey*, 'a scientific and
educational experiment' (Ward, 2012: 37). But how far was he prepared to
go to achieve those goals? We will never know, as he was evacuated from
the trek because of frostbite, which developed while he was fixing a ski
without his gloves on.

There is also a spiritual dimension to some of this extreme suffering. The
metaphor of Biblical suffering, linked with hell, is used by Fiennes in the
language he uses to describe his experiences – 'The following day was purga-
tory' (Fiennes, 1993: 164) – as well as his recital of lines from John Bunyan's
Pilgrim's Progress, which ran through his head during the journey: 'From the
place where he now stood the way was all along set so full of snares, traps,
gins and nets here, and so full of pits, pitfalls, deep holes and shelvings down
there that, had he a thousand souls, they had in reason been cast away'
(Fiennes, 1993: 206). There may also be a link to his hero Scott and almost
wanting to suffer to the same degree as a way of being like the explorer –
the ultimate example of walking in the footsteps of one's forbears

(Laing & Crouch, 2011; Seaton, 2002). Fiennes even compares his situation to Scott's at one point in the journey:

Only now, beyond the Pole, were we discovering the true meaning of *cold*. Our conditions in terms of body deterioration, slow starvation, inadequate clothing, wind chill temperature, altitude and even the day of the year, exactly matched those of Scott and his four companions as they came away from the Pole. (Fiennes, 1993: 161)

Another journey filled with pain is described by Best and Bowles with reference to their 800-kilometre walk along the Spanish Camino Way to Santiago de Compostela (Figure 2.1). They describe in excruciating detail injuries such as a knee which 'looks like a bright red water balloon filled to near

Figure 2.1 View of Santiago Cathedral
Source: J. Laing.

bursting. [Eli] has tendonitis, blisters and the haematoma on the back of her calf has turned black' (Best & Bowles, 2007: 182). She walks with a suspected fracture, too stubborn to have it X-rayed, lest it derail her pilgrimage. Colin's heel is a 'raw mess of weeping flesh' (Best & Bowles, 2007: 107). These are only their physical injuries. Mentally, they are spent, hitting rock-bottom at various points along the trail to Santiago de Compostela as they review their life and what they have made of it thus far. Eli and Colin cry and often fight, but still keep walking. Their journey becomes a vehicle for coming to terms with guilt, regret, bottled-up anger and despair; 'a focus on the self' (Norman, 2009: 66), which may be confronting and unexpected. There are no distractions along the Camino Way to blot out the emotional as well as physical pain, with the landscape often monotonous and their thoughts impossible to escape. By putting one foot in front of the other, endlessly, day after day, they both find a kind of redemption, and end their journey with a sense of peace and acceptance of what life has (and will) throw at them.

Overcoming obstacles

A subset of challenge is to overcome obstacles, in the vein of the Greek heroes who must prove themselves worthy of their adventures. Challenge may result from overcoming health issues such as illness, disease, an inherited condition or an accident of some kind and gives some explorer travellers something to prove. For Warren Macdonald, climbing after his accident was about 'expanding boundaries'. As he observes: 'I smiled to myself, knowing I'd just covered five kilometers of difficult terrain ... I knew now that I wouldn't be confined to the concrete footpath. *If I can do this already, who knows what is possible?*' (Macdonald, 1999: 162).

Jonathan, a polar trekker, described to us his childhood and adult problems with a congenital heart condition, which required surgery, and how this became something for him to overcome:

It actually put a monkey on my back that I felt like I had to prove that I wasn't less of a person, that I wasn't disabled. And so, as I'd take on each greater, more extreme challenge, to try to shake this monkey off, it's only recently I've realised ... 'you don't have to keep on proving yourself!' ... My life is documented as a person who has overcome challenges. That's true, I didn't choose that way, I didn't ask those adversities to come along, but they chose to come into my path and my path is set. It was passion that again allowed me to say, 'Well you're in the way, I'll have to go over you or around you but you're definitely not going to stop me'. So that's that.

Women explorer travellers, particularly in the 19th century, may seek to overcome prejudice and the limitations of society and their environment. Isabella Bird's journeys, for example, began when she was sent abroad for her health. Her illnesses did not stop her from engaging in intrepid and highly adventurous trips to places as far flung as China, Japan and the Rocky Mountains. In fact they seemed to cease once she was on the move (Murray, 2008), perhaps reflecting a restless spirit which found the conventions and restrictions of Victorian England suffocating. Certainly she thrived on isolation and danger, and her travelling saw her 'tapping into an innate lust for adventure and instinct for survival that would remain with her for the rest of her life' (Whybrow, 2003: 476).

Goal-setting and achievement

There is often a tendency to characterise explorer journeys as ego driven and narcissistic. Percy Fawcett was keen for his 1925 expedition not to be seen as a 'pampered exploration party, with an army of bearers, guides and cargo animals. Such top-heavy expeditions get nowhere; they linger on the fringe of civilization and bask in publicity' (quoted in Grann, 2009: 11). Yet early adventurous travels, such as those of the medieval explorer, were generally characterised as altruistic and ennobling (Leed, 1991). The journey was made for others, or for the sake of learning and discovery, which were not seen as selfish pursuits, but rather as commendable aims which honoured God or one's family.

The desire to turn adventures into perceived worthy endeavours might be argued to underpin the predilection of many modern explorers to use their expeditions as a platform for political or social activism. In recent times, media exposure is often used to focus attention on causes or deliver messages to the public. Expeditions raise money for charities, such as the treks to the North and South Pole by Prince Harry and various soldiers, in aid of Walking with the Wounded, which assists wounded service personnel. The existence of a personal cause behind a journey is not, however, a recent trend. In the late 19th century John Muir wrote books and articles that aimed to raise awareness of the need to preserve California's Sierra Nevada from development.

Some adventurers, like Polar trekker Catharine Hartley, are, however, honest about the fact that they had to be encouraged to raise money for charity, almost as an afterthought:

> Originally, I must admit, I had not really thought about raising money for a charity. It was hard enough finding £30,000, let alone more for

others. Then, one person in particular out of whom I was trying to prise money said, 'What you are attempting, Catharine, is great, but to be honest I think it's all rather self-indulgent. I mean, if you do succeed you will come home and achieve recognition but what does that do for anyone else?' I decided he was absolutely right. (Hartley, 2002: 61)

The charitable cause helps to assuage guilt over the privileged nature of the journey and the elitist connotations of being an explorer. Michael told us, using similar language to Catharine Hartley: 'With the trips, I give away all the money I make, over and above the very considerable cost of making the trips happen. It's a half a million-dollar exercise to go to the South Pole. And I wouldn't have it any other way because the trips are fairly self-indulgent things.'

There may also be scientific goals, which give the expedition a sense of authenticity and worth. The explorer is often at pains to make clear that they are making a contribution to human knowledge. Mary Kingsley gathered natural history specimens from her travels in Africa (Frank, 1986), which she discusses in detail in *Travels in West Africa* (1897). Even though Kingsley is clearly enamoured of the dangers and thrills inherent in risky journeys, which shines forth in her prose, she is also keen to give her journey a patina of gravitas and substance, which squares with Victorian notions of keeping busy and being useful to others. Kingsley therefore writes in the preface to her book that Dr Günther of the British Museum gave her 'the sense that the work was worth doing ... and sent me back to work again with the knowledge that if these things interested a man like him, it was a more than sufficient reason for me to go on collecting them' (Kingsley, 1897: xxiv). The African forests are akin to a library in her eyes, and she bestirred herself to learn its language, rather than just gaze at the 'pictures' (Kingsley, 1897: 200). Historian Tim Severin, in deciding to undertake the Brendan Voyage, in the footsteps of the Irish monk, argues, 'to warrant such risk and effort the endeavor had to produce worthwhile results. It had to strive toward a precise and serious purpose' (Severin, 1978: 10–11).

Those adventurers who cloaked themselves with academic scholarship were revered by others. Graham, a desert trekker, pointed out that his admiration of Thor Heyerdahl was based on the fact that 'he was doing something scientific'. Burke and Wills, albeit in a race to cross Australia and often criticised for their lack of credentials, also accumulated a great deal of scientific material, which has only recently been fully appreciated (Bonyhady, 1991). Space tourists are particularly keen for their exploits to be taken seriously, and often use scientific research as the key rationale for the huge amounts of money they spend pursuing their dreams (Olsen, 2005; Shuttleworth, 2002a).

Others look to inspire others or achieve social change. Nick Danziger (1988) rationalised his retracing of the Silk Road as a vehicle to promote international peace and understanding. Ross (see pp. 242–243 for full list of interviewees) wanted to use exposure from his expeditions to fight injustice: 'I see that since getting back from Siberia especially, I've seen what I do and the profile gained from what I do as a tool going some way towards enlightening people as to the issues facing the Berbers or in the case of Siberia, the reindeer herders.' Aaron, on the other hand, saw telling others about his travels as a social 'responsibility': 'I went with the premise that I'm not just doing it for myself but I'm also trying to [do] good for others, maybe expand their horizons a little.' Female travellers such as Robyn Davidson in *Tracks* (1980) mention the need to be a role model for other women, helping them break out of a defeatist mindset or see the possibilities in life. This mantle of being an ambassador for feminine achievements was, however, a surprise for some women. Sarah, for example, was astonished at the attention her Polar trek attracted from women, who comprised 'a lot of the audiences ... I guess that's alerted me to the fact that it is more significant for women to hear my story than it is for men'.

There is, however, clearly a *drive to be first* that infects some adventurers – to reach the highest peak, the source of the Nile, the geographic Pole. The race between Scott and Amundsen to the South Pole or between Fawcett and Rice to uncover the so-called Lost City of Z in the Amazon jungle (Grann, 2009) makes a powerfully dramatic narrative. Scott had contemplated in his diary the 'appalling possibility [of] the sight of the Norwegian flag forestalling ours' (Cherry-Garrard, 1922: 521), and the expeditionary team were devastated when this happened. They did not turn back, however, which suggests that being first was not the sole motive behind their journey. Wilson's diary entry insinuates that this was a driver for Amundsen: '[We] are all agreed that he can claim prior right to the Pole itself. He has beaten us in so far as he made a race of it. We have done what we came for all the same and as our programme was made out' (Cherry-Garrard, 1922: 522). Scott, however, acknowledges the disappointment: 'The Norwegians have forestalled us and are first at the Pole ... I am very sorry for my loyal companions ... All the daydreams must go; it will be a wearisome return' (Cherry-Garrard, 1922: 525).

Explorer travellers are so goal focused that they could almost be seen as *collecting* a series of goals, like a rare stamp or coin collection, wanting to 'complete the set' of challenges in remote places. Eric Weihenmayer (2001: 158) 'had a tick list of mountains a mile long', while Caroline Hamilton recalls wanting to go to Antarctica, but 'only so I could "do" all seven continents' (Hamilton, 2000: 24). There are a number of these goals that are common to explorer travellers, such as the Seven Summits, where mountain climbers attain the summits of the highest mountain on each continent (Weihenmayer, 2001). Butler (1996: 216) suggests that frontier destinations 'represent attractions to be acquired almost regardless of risk or hazard'. This collecting goal could also be seen as a form of *cultural capital* (Beedie & Hudson, 2003), where prestige or social distinction is viewed less in materialistic terms and more centred on exotic destinations or experiences.

Escape and freedom

The journey might represent freedom from societal constraints for some travellers (Kane & Tucker, 2004; Swarbrooke *et al.*, 2003; Wilson, 2004). Women like Isabella Bird and Edith Durham escaped the limitations of being a woman in 19th century England. The source material for Bird's *A Lady's Life in the Rocky Mountains* was a series of letters sent to her sister Henrietta. Wolff refers to Bird's 'dual life', with her 'masculine' travels contrasting with her sister's calm, domestic home life, and argues that many female explorers at the time saw their sisters as playing the role of 'conscience, home-self, and recipient of journal-letters' (Wolff, 1993: 233). Isabella Bird stopped travelling for a while when her sister died, and got married at nearly 50 years of age, perhaps unconsciously taking on her sister's more traditional role within society. She remained, however, a passionate advocate for a woman's freedom to travel, and refused her first invitation to address the Royal Geographical Society on the grounds that 'it seems scarcely consistent in a society which does not recognise the work of women to ask women to read a paper' (Murray, 2008: 238). She became the first female member of the Society in 1892, the year after her address.

Alexander Kinglake, an Englishman who travelled into Turkish Hungary in 1837, and then through the Middle East, was escaping Victorian mores about homosexuality, and sought a world with more fluid boundaries and a less judgemental view towards sexuality (Leed, 1991). He was disappointed in that his escape was from his own 'internalized prohibitions and repressions' which he took with him (Leed, 1991) and the world he encountered was unlike the fantasy he had created for himself. A retreat from modernity, and indeed his own self, was doomed to failure. Percy Fawcett found family life in Edwardian England stifling and sought a more thrilling existence. He wrote that he wanted 'to be ordinary', but 'deep down inside me a tiny voice was calling ... It was the voice of the wild places, and I knew that it was now part of me for ever'. Home was likened to a 'prison gate slowly but surely shutting me in'. Even the 'hell' of the Amazon was preferable and, what's more, he embraced and even 'loved' that netherworld, which held him in its 'fiendish grasp' (quoted in Grann, 2009: 101).

Escapism might be linked to life transitions. Graburn (1983) refers to *rite of passage* tourism, where a traveller is typically going through a period of change from one stage of life to the next, which makes this person look to the long term and an interest in self-development. Examples include a divorce or a change of career. For Matt Dickinson, climbing Mount Everest offered the chance to escape from personal problems at home: 'Everest, I began to realise, could give me the space I needed to sort out the mess' (Dickinson, 1998: 30). For Karen, trekking across Antarctica heralded change from her life as a forty-something mother:

> I just wanted to do something big, exciting, different and adventurous but I had no idea what it was. I thought about going overseas to work for a year, I thought about moving to the mountains . . . I just wanted to get out of the mundane and do something interesting, but I didn't know what. So I thought 'Oh well, I'm off'.

A number of the men we interviewed were frank about their need to get away from their everyday lives, including family. Rod articulated the strength of this drive to us:

> In fact it's a natural thing to get away from society and all of what society means and entails. And that includes the traffic around the cities, it includes sitting in front of the computer every day and responding to emails, but it [also] includes the dynamics of a family . . . You know, I love them to death and I would never want to be separated from them for good. I have a great relationship with my wife and my family and there are no issues about wanting to be away from them because I'm bored with them or they don't stimulate me enough. There's none of that, but it's healthy just to be separated from things that you do over and over again.

Self-discovery and self-actualisation

Travel for the purposes of self-development or self-awareness is not a new concept (Cronin, 2000; Rojek, 1993; Wearing, 2002), but is particularly pronounced for explorer travellers. When we asked for his overall assessment as to why people travel to far-off and adventurous places, Bryan was blunt: 'Personal growth. I think that's what it's all about.' Melinda con-curred: 'I would say the boundaries that we put on ourselves; our boundaries are not there [any more]. The limits of stress and how far you can go with something. I think everybody has a limit to what they can bear or what they can do, and I feel we went far beyond that in our headspace.' For Catharine Hartley, reaching the North Pole 'was the culminating point of an eighteen-year journey, both physical and mental, to become my own person. The patch of nothingness I stood on that day had a significance for me that reached back into the whole of my adult life' (Hartley, 2002: 2–3). Liv Arneson sees her treks as creative endeavours, which allow her the opportu-nity for self-expression:

> In some ways, asking me why I go to Antarctica is like asking a poet why she composes poems or a painter why he paints. I think human beings are all driven to create, to express certain aspects of ourselves in what we do ... Once you find that thing that makes you feel like the truest ver-sion of yourself, it makes more sense to ask how you could *not* pursue that thing. (Arneson & Bancroft, 2003: 20)

Self-actualisation can also be characterised by a desire to change or rein-vent oneself. Kira Salak, who travelled to the heart of Papua New Guinea, wasn't just looking to explore who she was. She also wanted to actively *change* who she was, through her travels to a place where no-one knew her (Salak, 2001). Harry, an adventurer who often travelled to live in different cultures, saw one of the chief values of his expeditions as the opportunity to become more vulnerable and open himself up to different experiences and other people:

> Exploration isn't necessarily about planting flags or conquering nature or mountains, and making your mark. I think we should try and get beyond that and veer to somehow the opposite, in other words opening yourself up and making yourself vulnerable and allowing the place to make its mark on you.

Harry talks about the need to 'peel back my layers as a Westerner'. Robyn Davidson used a similar metaphor:

> This desocializing process – the sloughing off, like a snakeskin, of the useless preoccupations and standards of the society I had left, and the growing of new ones that were more tuned to my present environment – was beginning to show. (Davidson, 1980: 201)

Spirituality and the sublimity of nature

Many explorer travellers allude to spiritual or transcendental motivations for seeking out risky and far-away places. Aaron's polar journey, for example, was an attempt, in his own words, to 'feed' his soul. He felt there was a missing part of himself that his expedition helped him to fill. Life at home, he felt, doesn't allow him to do this, filled as it is with commonplace and petty concerns. There is therefore a link here with escapist motivations.

Remote or isolated places are often associated with spiritual experiences (Laing & Crouch, 2009b). Anatoli Boukreev compares the 'transcendental summits' of mountains to 'the ruts of my everyday life' (Boukreev, 2001: 35), while the adventurer Wilfred Thesiger (1964: 1–2) writes of wanting to 'recapture the peace of mind I had known in the deserts of southern Arabia'. These days, that peace and isolation is often hard to find, in our lives of noise, manic busyness, multitasking and frantic attempts at connection and engagement with others through Facebook and Twitter. Robyn Davidson scoffs at the latter: 'It's such a false connection anyway. An empty connection. We're never really present. Back then, in the desert, I had a feeling of being profoundly connected, part of this phenomenon of life we're all in. And yet I was totally on my own' (quoted in Verghis, 2013: 5).

The adventurous experience is calming in its requirement for complete absorption and concentration, which attracts many adventurers. Doug likened it to a form of meditation: 'I mean you are very much focused on your immediate surroundings. Rather than doing it for thirty minutes or a six-day meditation camp, it's a pretty intense single-mindedness that one needs to achieve to succeed'.

Immersion in nature is seen as purifying the soul. Michael Asher (1988: 287) notes, as he and his wife completed their trek across the Sahara: 'Both of us felt sanctified after our wash in the Nile.' Robyn Davidson (1980: 154) also experienced this cleansing by nature during her desert journeys ('my mind was rinsed clean and sparkling and light'), which extends to her relationships:

> But strange things do happen when you trudge twenty miles a day, day after day, month after month. Things you only become conscious of in retrospect. For one thing I had remembered in minute and Technicolor detail everything that had ever happened in my past and all the people who had belonged there. ... People, faces, names, places, feelings, bits of knowledge, all waiting for inspection. It was a giant cleansing of all the garbage and muck that had accumulated in my brain, a gentle catharsis. And because of that, I suppose, I could see much more clearly into my

present relationships with people and with myself. And I was happy, there is simply no other word for it.

Liv Arneson is honest about the riskiness of her polar journeys but argues that this is outweighed by the tranquillity she finds in nature, helping her to understand herself and her place in the world: 'I am not a religious person, but the feeling is similar to the one I have heard some religious people describe: It's a reverence that makes everything else make sense. When I am in the wilderness, I know why I am here, what life is about, who I am' (Arneson & Bancroft, 2003: 20).

Honouring others

Sometimes the traveller wants to honour the memory of a family member or fulfil their dream. Caroline Hamilton was inspired to undertake her polar journeys by her father's influence, which had seeped into her consciousness from an early age. During her trek, she writes, 'I thought of my father. I so wanted to tell the others about him but I could not speak the words. Without him, I thought, none of us would be out on the ice at all, let alone at the Pole. He always wanted to explore the frozen wastes. I hoped he was proud of me, his only daughter fulfilling his dreams' (Hamilton, 2000: 3). Another team-mate of Hamilton's, Rosie Stancer, was also inspired by family influences and the desire to do something for another or in another's memory: 'Rosie was on a mission. That was clear. Her grandfather narrowly missed selection for Scott's expedition in 1911 and her husband William's grandfather was on Shackleton's in 1914. Rosie was getting there for both of them. And for herself. Her tough and feisty self' (Hamilton, 2000: 152). This focus on honouring others is also evident in re-enactments of explorations, which we explore more fully in Chapter 8.

The Explorer's Gene

In discussing the call to adventure, there are lots of diverse motivations that individuals have acknowledged in their post-journey accounts or which have been elicited by research into travel motivations (Laing & Crouch, 2009a, 2009b, 2011). Most of this work has focused on external influences like the media, role models or change agents like parents or teachers, or reading about the exploits of other travellers. These individuals also acknowledge internal drivers, like a healthy ego and a desire to be challenged. Adventurer and explorer Benedict Allen (2002: 771) alludes to 'dramatic advances' in our understanding of our genetic makeup as one of the 'frontiers of knowledge', which is yet to be fully explored. His comments are prescient.

The discovery of a mutated gene known as DRD4-7R, present in about 20% of the population, which helps control levels of dopamine, 'a chemical brain messenger important in learning and reward' (Dobbs, 2013: 50), suggests there might be a potential genetic basis behind exploration (see Crouch, 2013; Dobbs, 2013). Individuals with this gene variant are more likely to exhibit latent curiosity and a desire to take risks 'and generally embrace movement, change, adventure' (Dobbs, 2013: 50). Rather than labelling this 'the explorer's gene', it might be more accurate to see it as a factor *urging* people to explore, but not necessarily giving them 'the tools or traits that make exploration possible' (Dobbs, 2013: 51).

3   Preparation and Departure

MEN WANTED
for hazardous journey, small wages, bitter cold, long months of complete darkness,
constant danger, safe return doubtful, honor and recognition in case of success

This advertisement was allegedly placed in a 1914 English newspaper by the explorer Ernest Shackleton in order to recruit his expeditionary team for an assault on the Pole. Although possibly apocryphal (no copy has ever been found), it illustrates the dangers inherent in this type of activity and thus the importance of planning adventurous journeys down to the smallest detail. The riskiness of these endeavours is embraced and even courted, as the traveller encounters the 'unexpected, the new, the alien' (Leed, 1991: 40), yet ideally they are deliberate, carefully scheduled and highly organised journeys.

Time is required to seek adequate resources, carry out the requisite training and plan the travel, including logistics and a route. The right team needs to be selected, and funding may have to be sought from a variety of sources before the journey can take place. Timing of the travel is also a critical decision, and might involve an analysis of seasonal patterns and weather forecasts. The period leading up to departure is therefore a vital component of the success of the endeavour. A track record of disorganisation or failure may dissuade potential sponsors or team members from becoming involved in subsequent expeditions.

Leaving on a journey, however, is far more significant than just the nuts and bolts of logistics and planning. The departure is a *watershed moment*. Robyn Davidson felt that making the decision to depart on her expedition across the Outback had been her greatest challenge:

It struck me then that the most difficult thing had been the decision to act, the rest had been merely tenacity – and the fears were paper tigers. One really could act to change and control one's life; and the procedure, the process, was its own reward. (Davidson, 1980: 37)

Campbell (1949) identifies the departure as the first stage of the hero's journey, where destiny (and adventure) beckons. It can be seen as a symbolic *expulsion*, harking back to the Garden of Eden, 'a wandering away from origin so that knowledge as quest, curiosity, can be instituted' (Cronin, 2000: 119). Leed argues that it 'separates an individual from a fixed social matrix . . . and from that nest of relations that define identity' (Leed, 1991: 29). The traveller is cast adrift from *home*, and all that word connotes in terms of safety and security. This is of course part of the seduction of travel – it allows an escape from the everyday world, with its petty cares and absorptions (Cronin, 2000; Urry, 2002). As Burton comments:

Of the gladdest moments in human life, methinks, is the departure upon a distant journey into unknown lands. Shaking off with one mighty effort the fetters of Habit, the leaden weight of Routine, the cloak of many Cares and the slavery of Hope, one feels once more happy. The blood flows with the fast circulation of childhood. (Burton, 1872: 16–17)

Burton's reference to childhood is instructive. Leed observes that all departures are part of an endless cycle of separations, starting with the womb and ending in our death. The departure is 'both an ending and a beginning; it evokes a past and projects a future' (Leed, 1991: 31). It therefore has a primal quality, and we view departures with anxiety or excitement, but not usually indifference. This necessitates taking part in *rituals of departure*, which celebrate the significance and meaning of travel in our lives. In this chapter, we examine the importance of the departure, as the explorer traveller prepares for and then ventures forth from the known to the unknown world, including the rites that attend this setting forth.

Selecting the Team

The advertisement that Shackleton is said to have used to recruit for his Antarctic expedition has taken on iconic status, as it set the tone for the type of team the leader envisaged he needed for his assault on the Pole. A curious contrast is the 1860 advertisement lodged in newspapers by the Committee for the Victorian Exploring Expedition. They sought a gentleman to lead their expedition for the first north–south crossing of Australia and they were in direct competition with a rival venture organised by the adjoining colony of South Australia. Advertising in the paper got them 15 applicants, only four of whom had any experience. They plumped for police officer Robert O'Hara Burke, even though he had never been to the Outback (Murgatroyd, 2002).

Other recruitment processes were more serendipitous. Apsley Cherry-Garrard had met Bill Wilson, a member of Scott's 1901 expedition to Antarctica, at a shooting party in 1908, organised by Cherry-Garrard's cousin. He thrilled to the stories of bravado and courage told by the older man, and sought a position on the 1912 South Pole journey (Wheeler, 2003). Each of the team had a dedicated role, mostly as scientific experts (zoologists, meteorologists, biologists) or, in Cherry-Garrard's case, 'an adaptable helper', as Wilson dubbed him. There were limits in terms of the number of people who could be taken, and each member had to pull their weight and perform services that could not be done by others, yet be willing to muck in when the occasion demanded. Cherry-Garrard writes of the different roles of the crew: 'It is difficult to put a man down as performing any special job where each did so many' (Cherry-Garrard, 1922: 4). The sad coda to this tale is that Wilson, his inspiration for the voyage, later died alongside Scott. Both bodies were found by Cherry-Garrard, an experience that appears to have contributed to his lifelong depression and subsequent nervous breakdowns (Wheeler, 2003).

Any long-duration space voyage in the future, such as to Mars, will similarly have to grapple with the need to pare the number of participants to the bone, and ensure that expeditioners are multi-skilled and able to take on different roles and tasks as needed. The pilot is generally a dedicated specialist, but members of the crew might need to combine scientific expertise with other proficiencies, i.e. medical or engineering training. The Apollo 17

Figure 3.1 Harrison Schmitt, geologist and Apollo astronaut
Source: J. Laing.

mission was the first to take a dedicated scientist (Dr Harrison Schmitt, a geologist, see Figure 3.1), in order to assist with collecting the best samples possible. The astronauts who preceded him undertook geological training (Beattie, 2001) and were reportedly enthusiastic and diligent pupils. However, they were generally test pilots and this was not a substitute for having a geological expert on board (Schmitt, 2010).

The Planning Phase: Routes, Logistics and Training

Many explorers and adventurers have sought the guidance of organisations like the Royal Geographical Society (RGS) or the Explorers Club in their preparations for a journey. The RGS produced a guide, *Hints to Travellers*, which started out as Vol. xxiv of the Society's journal in 1854 and was later reprinted as a pamphlet. It covered everything from the packing of scientific instruments to the importance of careful observation and recording details for posterity (Driver, 2001; Grann, 2009; Murray, 2008). The RGS also held (and still does) seminars for explorers, covering various branches of science, including anthropology, geology and botany, as well as running examinations for fledgling adventurers who wanted the Society's imprimatur and blessing. Percy Fawcett argued that this thorough preparation 'bred me as an explorer' (quoted in Grann, 2009: 63). He took along the equipment recommended by the RGS on his Amazonian expeditions, but also often carried a copy of his favourite poem – Rudyard Kipling's *The Explorer* (Grann, 2009).

The diaries and log books meticulously inscribed mid-journey and the books produced after expeditions also form part of the body of knowledge that subsequent travellers can draw upon. Fawcett's leather-bound diaries are still kept by his family (Grann, 2009). Apsley Cherry-Garrard, one of the members of Scott's ill-fated trek to the Pole, observed that:

> The really important thing is that nothing of what is gained should be lost. It is one of the main objects of this book to hand on as complete a record as possible of the methods, equipment, food and weights used in Scott's Last Expedition for the use of future explorers. (Cherry-Garrard, 1922: xxxv)

Scott himself wrote 'The first object of writing an account of a Polar voyage [is] the guidance of future voyages; the first duty of the writer [is] to his successors' (Scott, 1905: xiii). The 1870 Washburn Expedition to Yellowstone was following up on stories gathered from illiterate fur trappers. It was only when the geysers and waterfalls were recorded in writing that they were regarded as really *discovered*. And it was these written accounts by educated *gentlemen* that convinced the US Congress to declare the area a national park (Langford, 1905). In contrast, the inability of Robert O'Hara Burke to keep any written account of his crossing of Australia has contributed greatly to the ridicule he has attracted over the years (Bonyhady, 1991; Murgatroyd, 2002).

Cherry-Garrard was emphatic about the importance of the planning phase, referring to Scott's first expedition in 1901–1904 as a disaster, where men had no idea how to use their stoves or lamps, erect their tents or put on their clothes in the freezing and life-threatening conditions. The calorie-rich pemmican was 'considered too rich to eat' (Cherry-Garrard, 1922: xxxiv), despite the fact that the severe cold was causing the men's bodies to literally eat themselves. These mistakes led to the death of one man, but it could easily have been worse. Scott himself admitted 'food, clothing, everything was wrong, the whole system was bad' (Scott, 1905: 202). These errors might be attributed in part to his dislike of the planning process: 'Scott used to say that the worst part of an expedition was over when the preparation was finished' (Cherry-Garrard, 1922: 1).

Scott's next expedition was, however, no better in terms of planning. Scott, Shackleton and their fellow adventurer Wilson showed signs of scurvy on their 1903 expedition to Antarctica, the result of inadequate food rations. On the other hand, Norwegian Fridtjof Nansen is usually held up as a model of expeditionary planning. He pioneered the light sledge and the use of dogs. It was the latter that was instrumental in his compatriot Amundsen's success in being the first expedition to reach the South Pole, beating Scott's team with their ponies.

Agatha Christie's archaeological expeditions with her husband, Max Mallowan, were also meticulously planned, as befits a writer for whom small detail was everything. The list of household utensils taken to a dig at Tell Arpachiyah in 1933 includes cooking equipment such as a frying basket, egg-beater, fish slice and frying pans, as well as the necessities for civilised dining such as tablecloths, napkins, finger bowls, salt and pepper shakers and a sauce boat (Trümpler, 1999a). The Mallowans always dressed for dinner, no matter how modest the surroundings. Max's precious archaeological texts filled numerous packing cases and were supplemented by grammar books (Turkish, Arabic and Sumerian), dictionaries (Arabic and English) and the Bible. Christie was careful to include pharmaceuticals such as aspirin, quinine and castor oil, some of which were used to help the local populace (Trümpler, 1999a). The records left of the Mallowans' preparations for a dig 'clearly illustrate the meticulous exactitude, forethought and experience that had to go into the preparations for even a relatively small dig' (Trümpler, 1999a: 171). Travel to Syria and Iraq in those days was mostly by rail. Christie notes that their huge quantities of luggage necessitated a Pullman carriage. Her husband would not have his books sent ahead separately, 'having once had registered luggage go astray' (Christie-Mallowan, 1946: 22) and was horrified when Christie entrusted their baggage to a porter when they arrived in Calais.

These days, the array of equipment available to the explorer traveller is overwhelming (Beedie & Hudson, 2003; Cater & Cloke, 2007; Williams & Soutar, 2005). Technological developments have led to new waterproof materials like Gore-Tex, and the array of boots now available is dazzling. Ingold (2004: 321) queries whether 'the technology of footwear [could be understood] ... as an effort to convert the imagined superiority of hands over feet, corresponding respectively to intelligence and instinct, or to reason and nature, into an experienced reality?' They also give the traveller a distinctive and authoritative gait (Ingold, 2004). Having the right gear is thus important, not just from a protective standpoint, but to give the explorer traveller the right *look* and *feel*. Shops like *The North Face, Helly Hansen* and *Mountain Designs* are a staple of suburban malls the world over, offering adventurers 'the real deal' and wannabes the vicarious thrill of pretending to dress like they are ready to climb Everest or trek through the Andes:

> There were rainbow-coloured tents and banana-hued kayaks and mauve mountain bikes and neon snowboards dangling from the ceilings and walls. Whole aisles were devoted to insect repellents, freeze-dried foods, lip balms and sunscreens... There was an area for 'adrenaline socks' and one for Techwick skivvies... It was as if the fewer the opportunities for genuine exploration, the greater the means were for anyone to attempt it, and the more baroque the ways – bungee cording, snow-boarding – that people found to replicate the sensation. (Grann, 2009: 64–65)

Many of the modern adventurers we interviewed relished the challenge of meticulous planning, even with the amount of labour-saving devices and technology available to them. Charlie calls it a 'really steep learning curve ... it's just like running a separate business', while for Keith, 'it's the most exciting thing I do ... I'm a good risk-manager and that's by asking lots of advice'. Graham enjoys 'interpreting the historical data [for his re-enactments of explorer journeys]. Then there's the challenge of organising the logistics, whether that be the food, the water, the permits'. Making sure one has the *correct* equipment is also important, and even becomes a form of authenticity. According to David Breashears:

> Ice climbing requires skills and technology all its own. For starters, you're using knife-sharp tools in the place of your fingers and toes. For the hand tools, specialized ice axes or hammers with slender, drooped ice picks. For the feet, the Foot Fang, a foot-sized platform, with teeth along the bottom and toes, which clips onto your boot instead of being strapped on like conventional crampons. Another Lowe Alpine systems invention. (Breashears, 1999: 141)

The hidden thread behind this quote appears to be that *real* climbers both pack the correct equipment and know how to use it in the right way. It is interesting to compare Breashears' comment with the desire of some explorer travellers for a deliberately low-tech expedition, also for reasons of maintaining the authenticity of the experience. For Robyn Davidson, taking a radio along on her trek through the Australian Outback was a sign of weakness: 'It didn't feel right. I didn't need it, didn't want to think of it sitting there, tempting me, didn't want that mental crutch, or physical link with the outside world' (Davidson, 1980: 97). She later bows to pressure, but calls it 'that big smudgy patch on the purity of my gesture' (Davidson, 1980: 124). Her reference to 'purity' is interesting – this is a common phrase evoked by explorer travellers to denote doing things authentically (Laing & Crouch, 2009b, 2009c). Michael Asher (1988: 9) refused to take any advanced technology on his Saharan desert sojourn, arguing that this 'would spoil the spirit of the adventure'. It might also despoil any re-enactment, where the traveller feels, like Roger Mear in his recreation of Scott's journey, that he or she needs to 'play the game by the old rules' (Mear & Swan, 1987: 234).

The challenge of adventurous travel may also be heightened by being preceded by a lengthy period of training or preparation, so that the individual prepares themselves for being worthy of encountering danger or the power of the wild. This might include working on physical fitness, improving map-reading or navigation skills and getting used to conditions likely to be faced on the journey, like lack of gravity or potential motion sickness for space tourists, extremes of temperatures for desert trekkers, or extreme cold or lack of oxygen for mountain climbers. Before his space flight, space tourist Mark Shuttleworth commented on the challenges he faced, including the physical challenges associated with training: 'I'm taking on one of the greatest challenges of my life . . . the actual training was incredibly disciplined, very detailed, very technical' (Shuttleworth, 2002b). Sean told us that he saw his pre-training for a flight as part of the authenticity of a potential space tourism experience: 'You feel you've actually done some proper astronaut training, so you [can] call yourself an astronaut properly, rather than just a passenger.' Richard described for us his preparation for trekking across the ice at the North Pole, and the necessity for a solo adventurer to be self-motivated:

> You always train for what you're going to be doing. So you try to make the training as realistic as possible. So for instance, if you're going to be pulling the sled, that's why I pull two truck tyres around a mountain. That's hard some days (laughs). That's the beauty of it too because you really require the biggest kick up the ass sometimes when you just think: 'Oh, I really can't be stuffed' … Yet I should be out there … So those things prepare you, they just harden you.

Clothing

The importance of clothing is constantly emphasised by explorer travellers, not just for safety, but also to look the part. Cater and Cloke (2007: 16) refer to this as a form of role play, part of 'performing adventure'. The theatre of exploration required adventurers to look the part. Fawcett was known for wearing his Stetson hat while exploring the Amazon jungle, but was also careful to pack clothing of 'lightweight, tear-proof gabardine. He had seen men die from the most innocuous-seeming oversight' (quoted in Grann, 2009: 7). Likewise, Henry Morton Stanley is often depicted wearing his 'exploring outfit of thigh-length boots and pith helmet' (Jeal, 2007: 151). Perhaps the most absurd was George Landells, initially the deputy of the Victorian Exploring Expedition. A horse dealer hired to source camels from India, he returned garbed in the red and white robes of a traditional Indian cameleer. Furthermore, he insisted on wearing this costume for the official departure (Murgatroyd, 2002).

Mary Kingsley is flippant about the advice she received from friends about medical preparations like quinine, mustard leaves and malt and cod-liver oil, but less forthcoming about the clothes she packs. Yet her Victorian tweed garb saves her skin in at least one scrape. Kingsley's fall onto a series of spikes in a game pit is cushioned by her 'good thick skirt', which she notes was more protective than the 'masculine garments' recommended by many people back home (Kingsley, 1897: 98). Proprieties are observed and Kingsley often bemoans the state of her clothes after a long trek. She holds up her trusty umbrella, even in a tropical deluge, observing that 'though hopeless it is the proper thing to do' (Kingsley, 1897: 221).

Putting on clothing can also be 'as much about stripping, metaphorically, as it is about the wearing' (Cater & Cloke, 2007: 16). The Victorian adventurer Isabella Bird (1879: vii) provides intensely practical advice in a note to her second edition of *A Lady's Life in the Rocky Mountains* on how to rough it as a woman, including riding cavalier fashion and the kinds of clothes that she wore on her travels 'for the benefit of other lady travellers':

> I wish to explain that my 'Hawaiian riding dress' is the 'American Lady's Mountain Dress', a half-fitting jacket, a skirt reaching to the ankles, and full Turkish trousers gathered into frills falling over the boots, – a thoroughly serviceable and feminine costume for mountaineering and other rough traveling, as in the Alps or any other part of the world.

Her garb is a symbolic paring away of the restraints inherent in being an upper-class female in Victorian England, just as Cater and Cloke (2007) suggest, which leaves her free to engage in experiences that most women of her generation and social class could only have dreamt of.

Other explorers use clothing as a disguise, a way to blend into the crowd but also to experiment with what it is like to *be* someone else. Cronin (2000: 153) notes that 'changing your clothes, your hair, your skin, learning a language, acquiring an accent, altering your habits – these have figured as means used by travellers to try out different personalities, [and] experience different (and familiar situations) from another point of view'. Disguise forms part of Richard Burton's legend (Phillips, 1999; Whybrow, 2003), the man who travelled to Mecca without his cover being blown, despite the fact he was 'a bold and arresting figure, six feet tall, with dark, glittering eyes' (Murray, 2008: 45). Burton 'grew his hair long, had a long beard and his face and hands, arms and feet were stained with henna. During the day he was probably taken for "a kind of Frank in a sort of Oriental dress"' (Murray, 2008: 52). He is precise in recording his apparel for the journey, which included a 'bright yellow cotton umbrella' (Laing & Frost, 2012: 117). In a section of his chapter 'I Leave Alexandria' in *Personal Narrative of a Pilgrimage to El-Medinah and Meccah* (1855) subtitled 'The Outfit', Burton provides advice to readers and would-be travellers through the Middle East on how to appear 'as a born believer':

> It is a great mistake to carry too few clothes, and those who travel as Orientals should always have at least one very grand suit for critical occasions. Throughout the East a badly dressed man is a pauper, and as in England, a pauper – unless he belongs to an order having a right to be poor – is a scoundrel. (Burton, 1855: 16)

The perfection of his disguise has been disputed by some, with Murray observing: 'Oriental mystery cuts both ways and it is possible that his survival may have been assured by the courteous discretion of his Arab hosts who did not wish to reveal that they had seen through his imposture' (Murray, 2008: 154) . His wife, who is acknowledged as adding to and keeping alive the Burton mythology after his death, stressed that Burton did not merely disguise himself with clothing, but effectively adopted a new rôle or persona during his travels (Murray, 2008). His skill at deception might also have extended to covering up his bisexuality (Phillips, 1999).

Like Stanley and Fawcett before her, Agatha Christie Mallowan (1946) was acutely aware of the need for practicality, but also the importance of *appearance* as a traveller to adventurous climes. She was as meticulous about her clothing preparations as she was with other elements of packing for an archaeological dig. Visiting the Tropical Department of an unnamed London department store, which consists largely of topees and jodhpurs, 'suitable wear for the wives of Empire Builders' (Christie Mallowan, 1946: 16), she notes that a shantung coat and skirt transform her 'into a Memsahib! I have certain qualms – but stifle them. After all, it is cool and practical and I *can* get into it'. Also on her list is a hat with a small brim, which won't fly off in strong winds, which she has made to her exact requirements. Four wrist watches are purchased, as Christie Mallowan knows from bitter experience that they are often provided to foremen on a dig, who tend to be as hard on the mechanics as the desert!

Sponsorship

Sponsorship is needed to fund an expedition, but generally requires a *quid pro quo*, the generation of publicity to highlight the sponsor's involvement. Patrick Woodhead remarks: 'Some people scoff when I tell them that raising sponsorship is undoubtedly the hardest part of an expedition. They picture climbing up mountains or traversing frozen wastelands to be unimaginably hard. In comparison to trying to prise money out of the tightly clenched fists of unexcitable executives, the odd mountain ridge is absolute child's play' (Woodhead, 2003: 52). Simon, a diver, also stressed the challenge and sense of achievement gained from successfully getting sponsors on board for an expedition:

> There's a great deal of excitement and anticipation about creating an idea, and then convincing other people that it is a good enough idea and worthy of them financing it. And there's a small sense of victory about that point.

Woodhead was consoled by the fact that the great explorers of the past have experienced similar challenges to raise funds, perhaps feeling a sense of *communitas* in the shared struggle and a sense of being in their footsteps: 'All the titans of polar travel have had enormous difficulty finding generous sponsors, no matter how big their name or how noble their cause' (Woodhead, 2003: 57).

Crossing the Threshold

The hero's journey depicted by Campbell (1949) characterises the departure as the crossing of the *threshold*. This is also mirrored in the *descent* of the ancient Greek *katabatic* narrative (Falconer, 2005; Frost & Laing, 2012; Laing & Frost, 2012). The threshold may be symbolic, a barrier in the traveller's mind that they must overcome in order to leave home. Examples include sorrow or guilt at leaving loved ones, the burden of domestic, work or filial responsibilities, or fear of the road ahead. Karen, a female Polar trekker, notes that this is often an issue for women: 'It's so easy to feel bogged down in domestic life. I'd love to encourage other women to get out and have a go at some big exciting things'. Leaving children behind is often the biggest wrench. Melinda, Karen's expedition partner, labels that 'the hardest part of the trip ... I had to leave them in [my parents'] care and I hated doing that because my parents are elderly.' She also notes that training had to be scheduled around her family's needs, and that it was generally easier for her male counterparts, who often had more back-up from a female partner.

While some of the men we interviewed expressed guilt at leaving young children, they usually prefaced that by emphasising that these journeys were necessary, important and non-negotiable. Richard tries to 'ignore' his family's concerns and observes how he doesn't 'want the worry, I just want the encouragement'. Graham argued that this desire to fly the domestic coop is important given the largely sedentary jobs held by many men in the 21st century, which don't 'allow you to use your intellect and your physical strength ... whereas guys in the 19th and 18th centuries, they had these wonderful careers doing stuff that was really intellectually challenging like exploring and mapping and navigating and also tremendously physically challenging'. His wife finds him 'the nicest I'm ever to live with' when he returns from an expedition, as 'I'm really relaxed and very calm'. The implication is that this need to be taxed, to push oneself to the limit, is hardwired in men, which is perhaps arguably the case for a certain proportion of *both* genders, given the discovery of a genetic predisposition towards exploration (see Chapter 2). In the case of the men we interviewed, explorer travel clearly contributes to their personal *identity*.

Some of these men did, however, feel uncomfortable with the knowledge that they live a privileged existence as an adventurer and often try to assuage this feeling by planning for and engaging in altruistic activities. As Harry notes:

> There is a certain amount of guilt I feel when I do my adventures, because I feel I'm getting maybe kudos, [a] certain income, not necessarily a great one, but certainly an income from what I do and the world isn't a play-ground ... the habitats that I've been in happen to be under threat. So I have a duty actually to be involved in some of these conservation move-ments and probably in fact should do far more than I do.

Another barrier delaying the departure may simply be a lack of resources. Ann Bancroft observes that, like guilt over family matters, this is often a major issue for women in particular:

> I was stopped for seven years from making a second attempt to cross Antarctica – because I was mired in debt ... So what I was out for this time around was not just to come up with a better model for expedition funding but to get access to solid funding and break the way for women who would come after me. (Arneson & Bancroft, 2003: 41–42)

A number of the people we interviewed mentioned the problems that Ernest Shackleton had in raising funds for his expeditions, and this seemed to make him a role model, inspiring them in their darkest moments to stay focused and resilient. Geoff, for example, noted his qualities of persistence and triumph over adversity, arguing that he felt a strong empathy with Shackleton and his financial struggles:

> I identify with Shackleton. [Firstly] because he had a lot of failures like I do, and also he struggled financially, which we all do in this environ-ment ... I admire his ability to, when he was so near to getting to the South Pole and winning the race, where he turned back because he knew they would die if they pushed on ... he knew that if he did make that, his name would be up there in lights and he would be financially secure for once, not in debt, because we're always in debt. I'm in debt now, overspent the last one, haven't been able to pay that off. So, I think the sheer courage of that one action, I just respect him so much.

This is perhaps an example of the traveller wishing to emulate and almost take on the persona of their 'hero', which has been labelled a *metempsychotic* journey (Laing & Crouch, 2011; Seaton, 2002).

The crossing of the threshold is also marked by leaving a distinct *physical location*, which might take on mythical qualities as a consequence, particularly if it is used often enough as a departure point. The RGS headquarters is a good example, discussed in the next section in terms of its role as a ritual of departure, or stepping on board a form of transport, such as a boat or train. Richard Burton left London for Southampton, to take a steamer, but chose to do so in disguise as a 'Persian Prince'. His departure was shrouded in mystery, which added to the theatricality and glamour of the explorer. The journalist Henry Stanley, on his way to Africa to find the whereabouts of the lost explorer David Livingstone, spent his last night in London in a hotel, to which he invited his mother and half-sister. She had abandoned him as a child, and he had spent part of his youth in an institution. Jeal (2007: 77) suggests that the act of inviting her to a grand hotel to bid him farewell as he set off on his great adventure was in part due to a realisation that he might not come back, but also was Stanley's way of asserting control over her and thus assuaging his childhood abandonment and powerlessness.

Rituals of Departure

A physical departure point is like a door, gate or portal, which symbolises the crossing of the threshold. Some explorers play this up for effect, selecting grandiose or public places from which to set forth, or a place that is highly significant, usually in its association with other explorers. A venue like the Royal Society or the RGS often gave an official imprimatur to an expedition and helped to create a formal *sending off* ceremony or ritual, which feeds into the explorer myth. On some occasions, the departure is so grand it cannot be held on a doorstep. When Burke and Wills departed from Melbourne, they did so from a public park, with 15,000 spectators in attendance. Still not fully organised, the explorers realised late in the day that they actually had to go through the performance of departure. Accordingly, as it started to get dark, Burke changed into his explorer's uniform, gave a speech and set off. They travelled about 3 kilometres before camping for the night and Burke slipped back into Melbourne to visit his girlfriend (Bonyhady, 1991; Murgatroyd, 2002).

Other explorers also allowed the general public the opportunity to cheer them on. The Duke of the Abruzzi, grandson of the first king of Italy, left Porta Nuova station in Turin in 1897 on an expedition to make the first ascent of Alaska's Mt Saint Elias. A hundred people turned up for the ceremony (Tenderini & Shandrick, 1997). His next expedition, to the North Pole, was even more fêted, with a naval salute fired by the military guns at the nearby fort as his boat slipped out of the harbour, alongside a private celebration hosted by King Umberto at the Quirinal Palace in Rome. Scott's *Terra Nova* was cheered off from Lyttleton Wharf in New Zealand, as Cherry-Garrard (1922: 47) expresses it, with 'telegrams from all parts of the world, special trains, all ships dressed, crowds and waving hands, steamers out to the Heads and a general hullabaloo'.

The symbolism of a formal departure from a significant place is still apparent with modern explorer travellers. The Pushing the Limits expedition, which supported a disabled adventurer, Andy Campbell, in his quest to circumnavigate the globe, departed from the RGS doors in 2012. They were photographed to mark the occasion alongside the RGS Vice-President and a representative of their sponsor, Land Rover. Fictional accounts of expeditions also used this convention to frame the hero's departure. Jules Verne in *Five Weeks in a Balloon* (1862) depicts his explorers attending a grand farewell dinner hosted by the RGS:

> Captain Pennet and his officers were also invited to this banquet, which was very lively and marked by many complimentary libations. Enough healths were drunk to assure to every guest the life of a centenarian ... During dessert a message arrived from the Queen. (Verne, 1862: 192)

Formed with the intention to 'help launch explorers on their travels' in order to 'chart every nook and cranny on the earth' (Grann, 2009: 50), the RGS became the arbiter of what was needed for an expedition and where these adventures should take place. Science became a deity to be served, and the RGS was fixated with the notion of collecting and documenting specimens, surveying, mapping and naming unexplored places, and solving conundrums like the source of the Nile.

Clubs also play their part in these rituals of departure. Phileas Fogg departs from the Reform Club in Verne's *Around the World in Eighty Days* (1873), and this was repeated by Michael Palin when he re-enacted the journey (Palin, 1989). Like the Societies, they offer a stamp of approval from one's peers and social class, and give the farewell an air of *gravitas*, as well as an element of levity, perhaps helping to block out last-minute nerves or moments of doubt. Palin refers to the columned and marble backdrop of the Reform Club as a fitting place for his departure: 'a place of consequence, grand and grave enough to add weight to any venture' (Palin, 1989: 13).

Some explorers, however, had reasons to keep their departure secret, to give themselves the greatest chance of reaching a goal first. Roald Amundsen did not tell his crew that he was taking them down to Antarctica for an assault on the South Pole until they were a month into their voyage.

The secrecy was partly the result of his impecunity – Amundsen needed to make a great splash with this expedition to raise much-needed funds – but also his recognition that success was more likely without the glare of the public spotlight. He later wrote: 'If at that juncture I had made my intention public, it would only have given occasion for a lot of newspaper discussion, and possibly have ended in the project being stifled at its birth. Everything had to be got ready quietly and calmly' (Amundsen, 1912: 59). This secrecy led to disgruntlement in some quarters, with suggestions that this was underhand and not *cricket* in the best British tradition. Sir Clements Markham, a former president of the RGS and one of the godfathers of Scott's son Peter (the other was James Barrie, the author of *Peter Pan*), was incensed, labelling it a 'dirty trick' in a letter to Sir John Scott Keltie, the then secretary of the RGS: 'If I was [Robert Falcon] Scott I would not let them land, but he is always too good-natured' (quoted in Crane, 2005: 382).

Amundsen's telegram, received by Scott in Melbourne ('*Madeira. Am going South. Amundsen*'), must have been a tremendous blow to the party. Cherry-Garrard was circumspect and perhaps generous after the fact, noting: '[Amundsen] had been in the Antarctic before Scott … and therefore did not consider the South Pole in any sense our property' (Cherry-Garrard, 1922: 40). But this leaves a question hanging – did Scott and his crew consider the South Pole *theirs* to conquer? While they acknowledged Amundsen's prodigious gifts and track record as an explorer, they perhaps did not realise what a threat he really was, how determined he was to succeed and how much heart and indomitable will he had: 'It meant, although we did not appreciate it at the time, that we were up against a very big man' (Cherry-Garrard, 1922: 41).

Saying Goodbye

The departure is a time of reflection and high emotion, even for seasoned adventurers. Farewelling loved ones can be difficult, as well as coming to terms with the realisation that one might not return. For Douglas Mawson, a letter had to suffice for his sweetheart Paquita, the woman who was to become his wife. In it, he writes:

I have a great longing to say something to you but I cannot in a letter communicate my feelings. You may be sure that I am going away this time far happier than last when there was no gem of priceless worth awaiting my return. You may be sure I will look after myself compatible with a dutiful endeavour to accomplish. (quoted in Robinson Flannery, 2000: 20)

In Mawson's case, there would be little opportunity to have his letters sent on to his fiancée. Some were received 22 months after they were sent.

For others, the departure might be a time of decision – of breaking off relationships or making proposals. Having left Melbourne, Robert Burke rode back to see the actress Julia Matthews. He presented her with a miniature portrait and asked her to marry him. She replied she could only give an answer when he returned. He also made his last will and testament, leaving everything to her. To leave one's money to an unmarried woman was such a shocking move in Victorian society that the expedition organisers resolved to keep it secret (Murgatroyd, 2002).

Many explorer travellers skip over this stage or only provide sketchy accounts of saying goodbye to friends and family in their accounts of their adventures. Perhaps this does not square with the explorer myth, and the ideal of the stoic hero or heroine, eager to leave home. Alternatively, some things might be too personal and private to share, in a world which is all too eager to discover all the nitty-gritty details of an expedition, and which paradoxically is encouraged to do so by an endless succession of explorer narratives and accounts that frame the journey.

4 The Journey

The journey can be characterised as the entering into a *liminal space*, where the explorer is freed from the shackles of everyday life, but is also exposed to dangers that are not present back home. These experiences often take a long time to unfold and develop and are the richer for it. Campbell (1949) conceptualises the hero's journey as encompassing a number of stages, one of which has been examined thus far – the Departure. The next stage is the *Trials and Victories of Initiation*, during which the hero suffers hardships and is tested during the journey. The 'physicality or embodied nature of travel' (Cronin, 2000: 133) is thus stressed in these narratives.

In this chapter, we focus on three aspects of initiation, which we argue are particularly pertinent to the explorer traveller – the *Road of Trials*, the *Apotheosis* and the *Ultimate Boon*. We start by examining the latter, as the goal or boon sought is often the driver behind the explorer's journey and frames their experience.

The Goal or Boon

Explorer journeys are often characterised by attainment of a goal or reward (see Chapter 2). What is sought after may differ among individuals, but this behaviour is characterised by a desire to prove something, to oneself or others, to take or seize opportunities and to endure hardship and setbacks. As Jack observed to us: 'You have to have a target; otherwise you just get sidetracked all the time.' These goals have to provide sufficient challenge so that they are not easily achievable. The sailor Jim Shekhdar notes: 'Objectives should be set just out of reach. One can have intermediate goals to give oneself a boost along the way, but the paramount goal must stretch the individual' (Shekhdar, 2001: 5).

Examples of goals or rewards include the honour of being the first to do something, the discovery of a new place, species or culture, the collection of scientific data, adding to the body of knowledge, or simply receiving the recognition of society (or those whose opinions are regarded as significant) for carrying out a momentous achievement. Iconic goals sought by explorers include seeking the source of the Nile or the Northwest Passage or climbing the highest mountain in the world. The race to be the first to climb Everest led to the loss of many lives, and the focus on reaching its summit still claims its victims, as is illustrated so viscerally in Jon Krakauer's *Into Thin Air* (1997). This can detract from the feats of endurance and personal growth that these experiences require, or the sheer enjoyment in pitting oneself against a challenging situation or surrounding oneself with the sublimity of nature. Ian Brown notes that Peter Treseder, his teammate on a trek to the South Pole, liked to 'execute some journeys in a particular style: you start, you finish and you spend as little time as possible in between' (Brown, 1999: 112). This approach put Treseder in conflict at times with his team because of his single-minded goal. Davis (2011) examines the historic conundrum over whether Mallory and Irvine reached the summit of Everest in 1924, interest in which has continued even after the bodies of the climbers were discovered in 1999 (see Anker & Roberts, 1999). He concludes that the question is surely academic and makes a mockery of their courage and suffering:

> They did, on that fateful day, climb higher than any human being before them, reaching heights that would not be attained again for nearly thirty years. That they were able to do so, given all they had endured, is surely achievement enough. (Davis, 2011: 573)

Obsessional goal setting

Some explorer travellers take goal setting and the desire for achievement to extreme lengths. Muir (2003: 203) describes this in a mountaineering context as 'summit fever'. Obsessional behaviour has been considered as an element of some travel experiences, including the desire to fulfil 'fanatical dreams' (Mackellar, 2006: 195). Celsi *et al.* (1993: 16) observe with respect to skydivers: 'an addiction model does appear to explain some of the motivation for skydiving', which relates to the 'high associated with the jump' and the constant need to re-experience the high. Part of the attraction may be the paradoxical feeling of control that this danger engenders in some people. Lyng calls this *edgework*, which 'may create more powerful feelings of competence than other types of skilful activities because it offers the right mix of skill and chance, a combination that maintains the illusion of controlling the seemingly uncontrollable' (Lyng, 1990: 872). There might also be elements of a *flow experience* here, where the skills possessed by the individual and the challenges posed by the activity are finely balanced, leading to an intense sense of enjoyment and pleasure (Csikszentmihalyi, 1975). These adventurous journeys can lead to a loss of self-consciousness, through intense experiences which demand total focus and absorption. In our interview, Bryan observed:

> That idea of losing yourself, becoming so absorbed in the action, that you lose any consciousness of self. And I think that has to happen with mountaineering, if you're going to do it. You're so focused on what's going on and that's one reason your fear disappears to a large extent. And when you can reach that stage, then you're OK, you know. You can do anything pretty well.

Obsessive goals can have deadly consequences for the explorer traveller. Krakauer acknowledges the link between these goals and the personality type that is drawn to mountaineering: 'Unfortunately, the sort of individual who is programmed to ignore personal distress and keep pushing for the top is frequently programmed to disregard signs of grave and imminent danger as well' (Krakauer, 1997: 185). Simon, in one of our interviews, made this stark observation about the perils of setting goals as a guide:

> What are you offering; what's the expectation? If you're saying: 'Hey, you give me lots of money and I'll make sure that you get there and back', you've made a deal with the Devil. And then *they* [the guides] have to wear the consequences. That's why Rob Hall ended up dying on top of the mountain [Everest]. He could have saved himself, no question, he was plenty experienced enough and he could have said: 'Sod you, I'm off.' He *had* to die up there because he had that transaction written in blood. He promised those people he'd get them up and back safely. If they weren't coming back, he wasn't coming back.

There can be an element of ego in the seeking of records or compulsive goal setting, which can lead in some cases to overblown pride and a false sense of omnipotence. Krakauer refers to Rob Hall's hubris as playing a role in his death:

> Hall had become so adept at running climbers of all abilities up and down Everest that he got a little cocky, perhaps. He'd bragged on more than one occasion that he could get almost any reasonably fit person to the summit, and his record seemed to support this. (Krakauer, 1997: 284)

Competition and the drive to be first

While the romantic view of the explorer is that they do it for *right* or *pure* reasons (Anker & Roberts, 1999; Heller, 2004), these journeys can become competitive, when one or more explorers are heading for the same goal. Burke and Wills were in a race with John Stuart to cross Australia (Murgatroyd, 2002). Scott and Amundsen battled it out for the prize of reaching the South Pole, with Peary and Cook engaged in a similar (but disputed) struggle for the North Pole (Conefrey & Jordan, 1998; Morris, 1909). Speke and Burton sought to be the first to find the source of the Nile (Allen, 2002; Jeal, 2011; Murray, 2008), while Laing and Clapperton raced to be the first to enter Timbuktu (Kryza, 2006). Percy Fawcett was bedevilled by a rival, Dr Alexander Hamilton Rice, who also sought the Lost City of Z – a mythical lost civilisation in the heart of the Amazon jungle. Unlike Fawcett, Rice had a private source of income, becoming a millionaire after marrying money (Grann, 2009). He had access to technology, such as planes that could land on water, radios and aerial cameras, and simply flew into the interior. Fawcett was livid. However, Rice did not succeed in his quest to find Z, although he became Professor of Geography at Harvard University and Director of the Harvard Institute of Geographical Exploration. Fawcett, on the other hand, failed to return from an expedition in 1925, and speculation continues as to his fate (Grann, 2009).

Inner rewards

Not all explorer travellers are driven by external goals or rewards. Some seek a more esoteric and personal outcome, linked to self-discovery and identity. Jonathan Waterman writes of his Arctic crossing: 'Any committed adventurer eventually learns that equipment and performance are just a means to that greater end of finding your place in the natural world' (Waterman, 2001: 11). Harry's travels in remote areas allowed him to be receptive to new experiences:

> Exploration is necessarily about planting flags or conquering nature or mountains, and making your mark. I think we should try to get beyond that and veer to somehow the opposite, in other words, opening yourself up and making yourself vulnerable and allowing the place to make its mark on you.

The emphasis on self-actualisation is illustrated by the following travel narrative. In this instance, the explorer traveller moves inward, and the sense of discovery is a personal one.

To the River (Olivia Laing, 2011)

Olivia Laing (like the explorer Alexander Laing, no relation to Jennifer!) subtitles her first book 'A Journey to the Surface'. Her use of the River Ouse in England as the route of her journey is also a metaphor for her reasons for travelling. After losing her job and the breakup of her relationship with her boyfriend, Laing needs to come to terms with these watershed moments in her life and rediscover herself through deep immersion in a tranquil yet slightly unsettling landscape:

> It was then that the idea of walking the river locked hold of me. I wanted to *clear out*, in all senses of the phrase, and I felt somewhere deep inside me that the river was where I needed to be ... What I had in mind was a survey or sounding, a way of catching and logging what a little patch of England looked like one midsummer week at the beginning of the twenty-first century. That's what I told people anyway. The truth was less easy to explain. I wanted somehow to get below the surface of the daily world, as a sleeper shrugs off the ordinary air and crests towards dreams. (Laing, 2011: 5–6)

This river is where the writer Virginia Woolf met her end – ironic (or perhaps not) given that the latter often used watery metaphors in her work. As Laing notes: '[Woolf] writes of *plunging, flooding, going under, being submerged* ... I wanted to know how these effortless plunges turned into a vanishing act of a far more sinister sort' (Laing, 2011: 10). While Laing is often thinking of Woolf during her journey, she is far away from the depressive state and breakdowns that led the elder writer to drown herself, and this is perhaps part of her fascination for Laing. She acknowledges that the 'nadir' of her despair occurs the night before she departs, when she telephones her ex-boyfriend and 'began to weep and found I couldn't stop'. By the time she leaves, Laing's healing had already begun: 'something in me started to lighten and lift' (Laing, 2011: 11).

The metaphors of water continue throughout the narrative. Laing is 'rinsed clean by sleep for the first time in months', the 'sun flooded by' and 'there are sights too beautiful to swallow' (Laing, 2011: 15, 30, 89). Watching the water or even trying to find its source is meditative. She remembers excerpts from poems; muses on buried aquifers, fossils, and the action of water on the landscape; dreamily watches the leaves float along with the current; and lies on the bank in the sunshine with her eyes shut, listening to the gentle sounds of nature, like bird calls and the hum of bees' wings. This is not to say the river is always a benign presence. Laing is aware that river sources 'are often freighted with taboos' (Laing, 2011: 21), like the Greek heroes who were blinded, or even killed, for seeing a goddess bathing, and Woolf's sad story is ever-present, as well as narratives like the soldiers who drowned fleeing Prince Edward's troops in 1264. Laing thinks about the river as a kind of *hell*, 'the notion of a world within our world, set deep, a world that can be entered only with difficulty by mortals' (Laing, 2011: 74), and again associates it with Greek mythology. Laing links her 'obsessive hydrophilia' (Laing, 2011: 56) to the books she has read – Dickens, Eliot, Hemingway and Twain – and her reference to Conrad evokes the river as a *heart of darkness*. The Ouse is thus a *cultural landscape* (Hardesty, 2003; Mayne, 2003; Reeves & McConville, 2011), incorporating both tangible and intangible layers of the past. It is replete with stories, literary associations and ghosts, as well as the built heritage left behind over the centuries.

Time stands still on this journey ('what country was I walking in, what age?') although paradoxically the march of days is relentless, like beats measured on a metronome. Laing generally does not talk to others she encounters, preferring to record snippets of their conversations, which are often aggressive, tension filled or banal – a far cry from the secret world she inhabits. Walking has slowed her down to the point where her mind has started to take flight and all sorts of strange ideas come into her head, as if she were dreaming. Small details are noted and treasured – the colour of poppies, the drift of pollen – but also the large questions such as whether there is life after death and the environmental catastrophes associated with human intervention in and inhabitation of the riverscape. The past keeps returning to haunt her ('Why does it linger instead of receding?'), but its ability to hurt her lessens, as 'the panic that had shadowed me for months dissolved away'.

Initiation and the Road of Trials

The inner journey depicted above involves overcoming self-imposed barriers and restrictions. In the traditional hero's journey, the trials are external, but lead to greater self-awareness and are often transformative (Celsi *et al.*, 1993; Laing & Frost, 2012). The traveller is tested, in a form of initiation, and must overcome hardships or ordeals, including extreme heat or cold, inclement weather, lack of food or water, difficult encounters with the Other, including language and cultural barriers (Cronin, 2000) and a perilous route. Sailor David Lewis tries to explain the allure of risk: 'none but the psychologically unbalanced are attracted to danger for its own sake. Risk is a disadvantage unfortunately often inherent in too many worthwhile ventures ... But [the explorer traveller] reduces any element of risk to an absolute minimum, acting out his destiny *in spite of* anxiety and fear' (Lewis, 1975: 50).

Communitas through shared experiences

In the hero's journey, a guide may be present from the beginning, or join the hero along the road of trials. They may play an important role in ensuring the survival of the traveller, through advice and assistance. The other important influence on the hero is that of the *companion*, also a traveller who shares the journey and who may provide comfort in a time of crisis. Don Quixote had Sancho Panza and Robinson Crusoe had Man Friday. Their chief contribution to the explorer journey is the creation of *communitas*. This is a concept often associated with the traditional pilgrimage (Turner & Turner, 1978). Arnould and Price (1993: 34) refer to it as 'a sense of communion ... feelings of linkage, of belonging, of group devotion to a transcendent goal'. This does not necessarily make the journey less personal. Even when travelling in a group and enjoying the benefit of comradeship, each individual is undergoing their own experience. As Rod notes: 'Yes, I'm travelling with one other guy, but in essence, we are autonomous, we are there by ourselves and we're independent.' For Virginia Morell, who travelled with a group to find the source of the Blue Nile: 'The different versions of our trip – what it was about and the events we'd experienced together – convinced me of one thing: We were all traveling down the Nile in our own Private Ethiopias' (Morell, 2001: 284).

A number of explorer travellers thrive on the experience of travelling or preparing for an expedition in a group or team. Karen enjoyed the *camaraderie* of working with others in the lead-up to her adventurous journey, as did Richard Branson (2002). Gyimóthy and Mykletun (2004: 871) note that adventure may involve 'a temporary suspension of social rules', allowing a CEO like Branson to work on equal footing with a mechanic as part of a team experience. Melinda found that working with a team on her polar trek offered her the challenge of getting to understand and deal with different personalities:

> You realise how different people are. Especially when you have to do a very close-knit thing together. To find the right sort of people is difficult, who have most of what you want within the working group, so it's interesting.

Mountain climbing appears to be particularly linked to these feelings of community. It is an activity that has strong networks, subcultures and fraternities among its exponents (Breashears, 1999; Krakauer, 1997) and this is an appealing part of the experience for many mountaineers. Murray refers to the shared narratives which bind climbers together: 'There's a lot of folklore surrounding it, there are a lot of tales and stories.' Andrew explains why he prefers climbing with a team:

> It's a very profound thing to spend two months risking your lives with other people; you get to know them pretty well. You have to get to know each other *really* well actually, especially at the high altitude because you tend to rely on each other without having to talk too much because even talking's hard work.

Occasionally the expected camaraderie does not materialise, perhaps where competitive natures take over. Helen found her attempt on Mount Everest disappointing for this reason: 'Here we are trying to climb the highest mountain on Earth, we've been living together all this time and no-one's working together.' Bear Grylls (2000: 238) labelled his Everest climb 'the most lonely work I had ever done'. This also echoes Krakauer's experience on Everest:

> In this god-forsaken place, I felt disconnected from the climbers around me – emotionally, spiritually, physically – to a degree I hadn't experienced on any previous expedition. We were a team in name only ... Each client was in it for himself or herself, pretty much. (Krakauer, 1997: 171)

Having a companion along might be confronting as much as comforting. The traveller invariably drops their mask in the face of adversity, leading others to see them as they really are, and not necessarily in a flattering light. Patrick Woodhead writes about how his journeys led him to expose himself, frailties, warts and all, to another person:

> Expeditions are always a great test of a person's friendship. There are invariably occasions when you feel so scared, or so exhausted, that there are simply no barriers left between you and another person. After days on a mountain, you can be at your lowest ebb, the bare bones of your character laid open for all to see. The simple truth is that you just have no strength left in which to conceal them. It is extremely scary to expose yourself like that, mainly due to the fact that you're not really sure how either you or, for that matter, the other person will react under such circumstances. (Woodhead, 2003: 8)

Travelling solo

A subset of explorer travellers do not take someone along with them, preferring the solo experience. Isolation can be an attraction for some individuals as well as the greater challenges involved in travelling on their own (Laing & Crouch, 2009b). Jon Bowermaster sailed to the Aleutians, and argues: 'Being free of human contact is one of the most compelling reasons for me to make the effort to journey this far from civilization' (Bowermaster, 2000: 243). Isolation makes travel more intense (Muir, 2003) and more authentic (Zurick, 1995). As Martin told us: 'It's just totally pure.' For Bryan, solo mountain journeys are more satisfying because 'social interaction can come between you and the environment. So if you remove that, you're much more focused on what's going on around you and that, for me, is important'. The way he describes his experience makes it close to *flow*, in the sense of total involvement or absorption in the moment (Celsi *et al.*, 1993; Csikszentmihalyi, 1975). Bryan also enjoys the freedom of not having to look after or rescue another human being on his solo climbs: 'I like to leave that [sense of responsibility] behind.'

Isabella Bird (1879) was ambivalent towards the isolation she experienced while travelling in the Rocky Mountains. At times she relishes the opportunity to be at one with nature: 'Its loneliness pleased me well. I did not see man, beast, or bird from the time I left Truckee till I returned' (Bird, 1879: 20). She is, however, a social creature, illustrated by her finely drawn portraits of the people she encounters and the way she is often embraced and befriended by others. Occasionally the remoteness of her journey overwhelms her. The intensity and neediness she feels towards travel, however, is paramount, exemplified by the comment she makes on seeing a group of wagons with oxen 'standing on their way to a distant part. Everything suggests a beyond' (Bird, 1879: 31).

Taking an animal along on the journey can ease loneliness and make the solo journey more palatable. Robyn Davidson was grief stricken when she lost her dog during her camel trek, as was Andrew Harper (1999):

> The day was long and silent. This was the first full day of the expedition when I had no one to talk to. Even though there were five of us, the camp had always been split in two – Mac and I on one team, the camels on the other. The whole tone of this trip has changed.

For Davidson and Harper, this loss was one of the great trials they endured in their Outback journeys. Cherry-Garrard similarly writes of the attachment Scott's expeditionary team formed with their ponies, which one by one had to be shot as the journey progressed. Bowers, in a letter home, writes about being 'fearfully cut up about my pony' who had to be killed when he couldn't get out of the icy water, and how Oates remarked: 'I shall be sick if I have to kill another horse like I did the last' (Cherry-Garrard, 1922: 152).

Encountering the Other

While some journeys to the periphery involve remote places without a local population (Antarctica, outer space), most involve some contact with an indigenous population: a phenomenon which has been labelled *transculturation*, defined as 'a phenomenon of the contact zone' (Pratt, 2008: 7). While an imperialist or colonial view of exploration might characterise these encounters as one-sided, with the power residing in the hands of the Western traveller, the reality was more complex. Many imperial explorers played 'an important role in the naming and thus "owning" of colonial territories' (Blunt, 1994: 32) and opened the way for colonisation. Yet there were also explorers who valued the opportunities to interact with and learn from local cultures, and even took up their causes upon their return in order to prevent some of the excesses of colonialism from continuing. Polezzi (2006: 171) argues that this narrative might in fact be empowering: 'bestowing visibility on people and events that would otherwise be easily forgotten'.

While David Livingstone was in Africa primarily for his missionary work, which was largely unsuccessful (Jeal, 1973; Murray, 2008), 'he recognised the value and complexity of the African culture' and his books 'contain a rich store of ethnographical as well as geographical information' (Frank, 1986: 31). He had attended meetings for the abolition of slavery and saw the rise of commerce in the African continent as the way to stop this. His tolerance had its bounds, however. He writes of 'an intense disgust of paganism' and the 'degradation' that he encounters among the indigenous population. He is keen to create 'a centre of civilization' in Africa, and his exploration was partly aimed at finding a suitable site for this purpose (Murray, 2008). Yet he also writes of acts of kindness by locals, such as the chief's son who gives up his blanket to ease Livingstone's discomfort during a storm (Murray, 2008). Percy Fawcett also made genuine efforts to befriend the indigenous people he met on his Amazonian expeditions, although it appears that he slipped up once, ordering his fellow expeditioners to fire on some Indians in Bolivia, according to one of their accounts, but 'was apparently so mortified that he doctored his official reports to the RGS and concealed the truth his entire life' (Grann, 2009: 131).

These individuals, however open to new experiences and different cultures, are still firmly the products of their age. The modern reader of their accounts can be jarred by comments that these days would be regarded as racist or patronising (Frank, 1986). This should not take away from their achievements in opening up our minds to worlds beyond the West, and the political and social reforms they managed to influence. Mary Kingsley (1897: 147) wrote that studying the 'African form of thought' was her chief reason for visiting West Africa. She is at pains to be seen to be open minded, arguing: 'The study of natural phenomena knocks the bottom out of any man's conceit if it is done honestly and not by selecting only those facts that fit in with his pre-conceived or ingrafted notions' (Kingsley, 1897: 154). Kingsley explores these civilisations, and in so doing, is humbled, noting how the conceit was taken out of her in realising how little she knew and how much of her knowledge was dependent on her upbringing and the familiarity of conditions back home. Occasionally, however, her worldview makes one cringe. There are comments about 'wild, wicked looking savages' (Kingsley, 1897: 83), and she labels her guides Grey Shirt, Singlet and Pagan 'for their honourable names are awfully alike when you do hear them' (Kingsley, 1897: 75).

Yet on her return to England, she set about trying to save Africa and its people 'from the damage she felt was being wrought by missionaries and the British government' (Frank, 1986: xxi). Kingsley supported polygamy in Africa, in opposition to the missionaries, because she saw how it 'governs the daily life of the native' (Kingsley, 1897: 63) and had practical benefits, such as dividing up the work between wives and allowing women to recover from childbirth without the need to resume sexual relations, and thus avoid a series of quick pregnancies in succession. Interfering with the social fabric risked creating 'a moral mess of the first water all round' (Kingsley, 1897: 64). Her views on political rule, outlined in her second book, *West African Studies* (1899), were also uncharacteristic of her age. She did not espouse self-government of Africa, but neither did she want to see the European governments in control. Instead, she thought that traders should take charge, as this would offer the greatest protection to the people and their culture (Blunt, 1994; Frank, 1986). In this way, her views were aligned with those of David Livingstone.

The horror, the horror

Visceral tales abound of explorers pitting themselves against the elements and many make uncomfortable reading, which is arguably part of their intense fascination. Cherry-Garrard, writing about his winter journey in Antarctica, acknowledges this phenomenon, with his comment, 'Yes! comfortable, warm reader', which precedes his observation: 'Men do not fear death, they fear the pain of dying' (Cherry-Garrard, 1922: 287). The readers feel they are part of the story, but also feel somewhat guilty at being in relative comfort, while being regaled with tales of the dreadful hardship suffered by the author. In this way, the vicarious pleasure of reading adventure narratives is akin to the enjoyment of crime fiction (Laing & Frost, 2012).

James Murray, one of Percy Fawcett's expeditionary team on a trek through the Amazon in 1911, survived Antarctic expeditions with the likes of Shackleton, but found Fawcett's idea of adventuring unendurable. He had maggots growing inside him, never had enough to eat, and found it impossible to sleep from 'bites, cold and tiredness' (quoted in Grann, 2009: 113). Fawcett labelled him a 'weakling' and refused to take more frequent rests, believing that others could keep up the same relentless pace that he set himself and share the same rude health. Murray left the expedition and the others assumed he would die before he reached civilisation. He surprised them all by clinging to life, and came home full of anger at his treatment by the expedition leader. His story has a sad ending – he took part in an Arctic expedition in 1913, and joined a small party of mutineers escaping from their ice-bound boat, all of whom were subsequently lost without trace (Grann, 2009).

The trials or hardships are seen as an integral part of most explorer journeys. Jim Shekhdar observes: 'Of course, it was going to be dangerous. If the adventure was going to be worthwhile, I needed, almost by definition, to step as close to the edge of the abyss as I dared' (Shekhdar, 2001: 193). Gyimóthy and Mykletun (2004: 859) deconstruct the risky character of *deep play*, exemplified in Arctic trekking, as 'an activity in which the stakes are so high that it is irrational for anyone to engage in it at all, since the marginal utility of what one stands to gain is grossly outweighed by the disutility that one stands to lose'. Many explorer travellers beg to differ about what is being gained. Patrick Woodhead describes why Ranulph Fiennes seeks out risk at the poles:

> By deliberately subjecting himself to the mercy of the elements, Fiennes was able to feel more basic, primeval emotions, which, quite simply, made him feel more alive. It sounds a little clichéd, but it was a convincing argument – people don't head off to Antarctica to avoid risk, they go there because risk is exactly what they seek. (Woodhead, 2003: 19)

Dangers faced may offer opportunities for self-actualisation. Free diver Pipín Ferraras (2004: 172) writes: 'I discovered something new about myself each time I went down.' He sees the danger as an important element of that: 'Pushing the envelope. Going to the edge. Finding out what you were capable of. Testing the inner you' (Ferraras, 2004: 68).

This stage of the hero's journey is closely aligned to the Greek mythological construct of the *katabasis* or descent (Falconer, 2005; Holtsmark, 2001). This is literally the journey to hell, where the individual undergoes hardship, and is transformed: 'Through suffering, the traveller learns what they are capable of and understands themselves more deeply' (Frost & Laing,

2012: 215). The central motif of these journeys is therefore *identity* (Frost & Laing, 2012; Holtsmark, 2001). We argue that the katabatic narrative frames many travel narratives such as books and films, and underpins our imaginings about travel (Laing & Frost, 2012; see also Chapter 6). While we focus in this book on Campbell's (1949) hero's journey as the framework for our discussion, we are cognisant of the value of the *katabasis* in the context of adventure travel, particularly in attempting to understand the appeal of the nightmarish aspects of explorer journeys.

One of the more famous *katabatic* journeys is documented by the young adventurer, Apsley Cherry-Garrard, who accompanied Scott on his doomed attempt to reach the South Pole in 1912.

The Worst Journey in the World (Apsley Cherry-Garrard, 1922)

We focus here on his section titled 'The Winter Journey', as this offers some of the most searingly honest prose in adventure writing and is both inspiring and alarming in its account of extreme suffering. Cherry-Garrard was asked along on Scott's 1912 expedition to reach the South Pole to take part in a scientific sub-expedition that sought to collect the eggs of the Emperor penguin and study the embryos, which 'may prove the missing link between birds and the reptiles from which birds may have sprung' (Cherry-Garrard, 1922: 240). The three men, Bowers (Birdie), Wilson (Bill) and Cherry-Garrard were forced to travel to the penguin's nesting area in the harsh darkness of July, as this was when it was believed that the penguins laid their eggs, and their story is a tribute to indomitable will and strength of character.

There are so many things that test this small group, that it would take much of our book to outline them. Suffice to say that they had to get used to working largely in the dark; it took them four hours each morning to get into their harnesses (due to the build-up of ice overnight with their breath and sweat); they could hardly sleep; and all of Cherry-Garrard's teeth 'split to pieces' (Cherry-Garrard, 1922: 298). Their tent blows away in a hurricane-like blizzard (but is miraculously retrieved, which essentially saves their lives). Cherry-Garrard's spectacles ice up and become useless, requiring him in many instances to carry on blind, which led to two of the precious five eggs being smashed on their return from the beach. They fall into numerous crevasses and have to be hauled out by their exhausted teammates. This was a living hell:

> The horror of the nineteen days it took us to travel from Cape Evans to Cape Crozier would have to be re-experienced to be appreciated; and any one would be a fool who went again: it is not possible to describe it.

The weeks which followed them were comparative bliss, not because later our conditions were better – they were far worse – but because we were callous. I for one had come to that point of suffering at which I did not really care if only I could die without much pain. They talk of the heroism of the dying – they little know – it would be so easy to die, a dose of morphia, a friendly crevasse, and blissful sleep. The trouble is to go on. (Cherry-Garrard, 1922: 242)

Cherry-Garrard emphasises how the three men did not lose their humanity in the midst of this wretched existence, which was the true heroism of the journey. He often wanted to 'howl', but instead invented a mantra to help him continue on – 'You've got it in the neck – stick it – stick it'. The constant dangers faced meant that it was not possible to let his mind wander off on reveries of the past or the future. Instead, he had to 'live only for the job of the moment' (Cherry-Garrard, 1922: 248). His companions never bowed down to despair, nor ill-humour, understandable though that would have been. Cherry-Garrard (1922: 251) calls them 'gold, pure, shining, unalloyed'. This was the best of British fortitude, and an exemplar of the *keep calm and carry on* mentality. The terrible denouement for Cherry-Garrard (1922: 251) was that 'these two men went through the Winter Journey and lived: later they went through the Polar Journey and died'. Cherry-Garrard found their bodies and the horror never left him. He later dedicated his book as a memorial to his two fallen comrades.

Another postscript involves their reasons for undergoing these trials. Cherry-Garrard makes it clear that for most of his expeditionary team, it was not 'money or fame'. It was to extend the frontiers of knowledge: 'There was an ideal in front of and behind this work ... We travelled for Science' (Cherry-Garrard, 1922: 231–232). The seriousness of this purpose made it possible for them to deal with the pain, the danger and the discomfort, although Wilson tells the others that he had not known how bad it was going to be and feels guilt at involving them in his quest. Scott himself had tried twice to dissuade him. The irony was that when they returned, the initial response to their specimens was lukewarm, and Cherry-Garrard is bitter at the indifference and rudeness of the scientist at the Natural History Museum, who seems to dismiss both him and his efforts as inconsequential, as well as trivialising the deaths of his teammates. It takes time before the eggs are regarded as having some scientific validity. Professor Ewart of Edinburgh University writes in Cherry-Garrard's book: 'If the conclusions arrived at with the help of the Emperor Penguin embryos about the origin of feathers are justified, the worst journey in the world in the interest of science was not made in vain' (Cherry-Garrard, 1922: 310).

Playing at danger

Although the risks inherent in these journeys are real, there is sometimes a sense that the explorer traveller is enjoying the excitement of toying with danger, seeing it almost as a game to be won. This is illustrated by the metaphors or imagery associated with games or gambling that many of them use when describing their experiences (Laing & Crouch, 2009c). Karen refers to her South Pole trek as 'this gigantic jigsaw', while Bear Grylls (2000: 111–112) views his climb up Everest as 'a giant game of snakes and ladders, and like in the game, the higher you go, the further you have to fall'. Jim Shekhdar's solo row across the Pacific is successful, and he notes with relief:

> I had made it. I had made it, alone and unaided. In a sense, I had rolled the dice and won. Maybe it is the thought of what would have happened if I had gambled and lost that overwhelms me when I talk about those memorable scenes on the beach. (Shekhdar, 2001: 237)

Others try to justify their behaviour by ignoring the risks or arguing that their experiences are less dangerous than other more socially acceptable pursuits. Mark Shuttleworth (2002c) doesn't 'dwell on morbid thoughts', while Keith observes: 'The riskiest thing I do is to drive my car occasionally into the city.' Csikszentmihalyi (1975: 84) notes in his study of climbers that there is an 'intriguing recurrence of the statement that rock climbing is less dangerous than everyday activities, such as driving a car or walking down a street'. Playing down the risks of what one does might in some cases be a form of humility, linked to embarrassment at the sacrificial or obsessional side of explorer travel. A number of these individuals, for example, described their experiences to us as *fun*. There might also be an egotistical element here, a desire to appear nonchalant in the face of extreme danger, which makes what is achieved by the explorer traveller seem even more remarkable (Murray, 2008; Thompson, 2011). This is often a British trait, associated with maintaining the stiff upper lip but also the use of self-deprecating humour (Murray, 2008). Women in particular use this as a device, notably Mary Kingsley, Isabella Bird and Dervla Murphy, perhaps to avoid censure as a female traveller exposing herself to a dangerous situation (O'Neill, 2009). Its use as a literary device in Eric Newby's *A Short Walk in the Hindu Kush* (1958) is discussed later in this chapter.

Weighing the risks

Some adventurers refer to taking what they term *acceptable risks*. Simon weighs up whether the dangers he takes are worth the effort: 'A lot of these things, ultimately the risk you are taking is that you are going to lose your life. Now is whatever you're going to see worth risking your life for?' Liv Arneson notes: 'A good explorer challenges the limits of what is: and yet, a good explorer also knows when to quit – when the risks are too great or the efforts will be wasted. The line between tenacious and foolish is very thin' (Arneson & Bancroft, 2003: 171). For Doug, this means that 'when it's wrong, I know it's wrong [and] I'll turn around'. Some have scaled back their activities with age and greater experience. In his interview, Murray observed:

> I have a very clear view or idea in my head and my mind of the level of risk that I'm prepared to accept. And I know people, people I climb with, some of them are prepared to accept more risk and some of them are prepared to accept less risk but as I get older and I'm more careful and very selective about the people I climb with, they generally seem to be on a similar sort of level with it, with that decision-making process. And if I consider it's too dangerous, then I'll quite happily walk away from it basically.

Another explorer we interviewed was Martin, who gave up solo climbing because of the danger, which he was not prepared to undergo as a parent:

> I do enjoy solo climbing but I don't do it now because it's simply too dangerous. It's something I did when I didn't have dependants but the need for me to do something like that is nothing compared to the need for me to be around for my kids.

Maintaining morale

Explorer travellers use many devices to keep their spirits high. Literature can be a source of motivation *during* as well as leading up to adventure travel, allowing a brief diversion from the slog or hardship. Jon Muir brought along a book on his solo traverse of Australia and found it inspirational reading: 'Dark clouds and a little spitting rain; time to do a bit of reading ... It is the journal of [Hubert's] crossing of Antarctica and sounds like the escape I need' (Muir, 2003: 21). Exploring Yellowstone, Nathaniel Langford (1905) recited Lord Byron to his comrades. Cherry-Garrard's team during the Winter Journey sang hymns and songs, while in the hut before they depart, the men celebrated mid-winter, with a decorated Christmas tree, jokes, whistles and pudding. The use of rituals helps the men deal with boredom and confinement, and marks the passing of time, much as they do these days in the modern Antarctic station (Suedfeld & Steel, 2000; Wood *et al.*, 2000).

The Victorian female explorer was particularly adept at keeping her chin up during dark moments on a journey. Mary Kingsley often uses humour to alleviate the tension in a situation, and writes: 'There is nothing like entering into the spirit of a thing like this if you mean to enjoy it, and after all that's the wisest thing to do out here, for there's nothing between enjoying it and dying of it' (Kingsley, 1897: 140). Isabella Bird is equally able to laugh at herself, describing how Mountain Jim 'dragged me up, like a bale of goods, by sheer force of muscle' up Long's Peak (Bird, 1879: 109). This sense of the ridiculous no doubt sustained her in difficult moments. It is hard not to admire her and warm to her relentless indomitability and practical nature, which sees her cope with conditions that would horrify many of her compatriots.

This use of humour to diffuse tension but also paradoxically to place in sharp relief the dangers in which the explorer finds themselves, has become almost an affectation, as is illustrated by the following well-known travel book, which 'plays to a nostalgic, middle-class audience' (Holland & Huggan, 2000: 45).

A Short Walk in the Hindu Kush (Eric Newby, 1958)

Eric Newby is a failure in the fashion business, which he exaggerates for comic effect. Going on an expedition is a way of escaping from rude agents and uninterested buyers. His friend Hugh had written to Newby about a mountain he had tried to climb and how he tried to enter Nuristan but did not get permission. This is similar to Fred Burnaby's (1877) reasoning for riding to Khiva – the idea that an Englishman should be able to go wherever he wants, without restriction. Newby's account contributes to the myth 'of the English gentleman abroad' (Holland & Huggan, 2000: 23), the journeys of the privileged patrician (Thompson, 2011).

Newby deliberately plays up his amateur status as a mountain climber, which Holland and Huggan (2000) argue is one of the three themes dominant in his writing. By doing so, he paradoxically makes what he has done even more meritorious, as it has been achieved despite a lack of training and expert knowledge. The other two themes are intertwined with the first – 'anachronism and imposture' (Holland & Huggan, 2000). Newby is the last remnant of a dying breed – the non-professional who rises to meet a challenge – rather like the character of Rudolf Rassendyll in Hope's *The Prisoner of Zenda* (1894). He is pretending to be clumsy, awkward and inept, the epitome of the Hooray Henry dreaming of drinking Pimms while climbing, but in reality his travel through the Hindu Kush demonstrates his fortitude and moral fibre.

The title of the book is itself ironic and self-deprecatory – a 'short walk' that nearly had him killed at different stages and required him to overcome hunger, cold, pain and fatigue. Holland and Huggan (2000) argue that his work is a form of *camp*. Even the list of provisions taken on the expedition is used to highlight his bumbling efforts. Newby has never heard of some of the items on the list of equipment he is given, and takes the wrong boots, which leave his feet raw and bleeding. One wonders whether he packed his brolly and a bowler hat. It strains credibility that a man who survived being a prisoner of war in Italy during WWII and escaped over the Apennines, would be quite so naive, but this is the fiction that Newby has set up.

One laughs at his descriptions of meals where the highlight is spoonfuls of jam from a tin, and meeting a shepherd 'wearing long robes and a skull cap, the image of Alec Guinness disguised as a Cardinal' (Newby, 1958: 164). Yet he notes in a moment of self-awareness that his jests are the tedious type 'that explorers employ to keep up their spirits' (Newby, 1958: 94) and later on wonders 'why I had become an explorer' (Newby, 1958: 239). Newby subsequently meets with Wilfred Thesiger, author of *Arabian Sands* (1959), who welcomes Eric and Hugh as *fellow explorers*, yet comments that they must be 'pansies' (Newby, 1958: 248) for sleeping on air-beds. The photos Newby uses in his book, however, are not humorous – they show the group sleeping rough and climbing rugged terrain. Occasionally, he mentions the atrocities he encounters, including the dead body of a young traveller with a 'skull smashed to pulp' (Newby, 1958: 129). Newby is clearly aware that what he is doing is no less than *exploration*, regardless of his levity and mocking tone.

The Apotheosis

Once the trials are over, the hero reaches *apotheosis*, a divine state where fear has been eliminated and their full potential has been reached (Campbell, 1949). This stage of the hero's journey may be accompanied by intense emotions, and represents a *catharsis*, a term drawn from classical Greek drama which represents 'an emotive release ... a figurative purification or "cleansing of emotions"' (Celsi et al., 1993: 3). It has been argued that a dramatic worldview plays a role in high-risk leisure consumption like sky-diving (Celsi et al., 1993), and frontier (explorer) travel (Laing & Crouch, 2009c). Adventure is thus undertaken 'within a framework of myth and dramatic story line' (Trauer, 2006: 185), which Cater and Cloke (2007: 15) label *performing adventure*. Ness (2012: 120), in the context of rock climbing in Yosemite, similarly refers to a 'space of performance'. There are numerous examples of theatrical imagery used with respect to adventure travel. Worsley (2011) refers to the 'cast' in his re-enactment of Shackleton's journey, while Evan observed to us that he enjoys the 'theatre' of ballooning. Benedict Allen uses these metaphors when writing about his trek through the Australian Outback: 'Once, years ago at school, we had done a play, Beckett's *Waiting for Godot*. This camp had the same emptiness as the set, the inaction of the characters' (Allen, 1992: 282). There is also a potential nexus here with adventure as a form of *deep play* (Gyimóthy & Mykletun, 2004).

The apotheosis can be illustrated by the deeply spiritual experiences that some explorer travellers have in nature, which may enrich lives and lead to a deeper self-knowledge.

The sanctity and sublimity of nature

The search for sublime natural experiences might be characterised as an illustration of the romantic gaze at work (Urry, 2002: 78). The traveller's desire to be alone and at one with nature often forces them on an endless quest for seclusion and sublime natural scenery, and nature is seen as awe inspiring, dwarfing the presence of humankind (Squire, 1988; Waitt *et al.*, 2003). Cohen (2004: 322) refers to 'the existential authenticity engendered by total involvement in the tasks of the trip and the awe-inspiring sublimity of the surroundings'. For example, Geoff told us: 'I love being at sea so for me, it is definitely sort of a spiritual environment for me ... Even in the typhoons I found it amazing to see nature in full force like that – it's moving.'

This transcendence can be shared through travellers' accounts, some of which are lyrical and highly personal. Mountaineer David Breashears once glued a flake of rock back on a cliff-side, and justified this action by explaining, 'I felt like I had desecrated a shrine' (Breashears, 1999: 58). He later explains the spiritual pull of the mountains by relating another climbing experience:

> Near the top I pulled from the shadows into the sunlight and something peculiar happened: I felt lifted from my darkness. It was a profound feeling that would eventually fade but, for a brief moment, the chaos in my soul subsided. In the years to come I would find the same measure of spiritual relief and self-awareness at other moments on the world's mountaintops, in havens of light. (Breashears, 1999: 68)

Referring to 'that light' Breashears (1999: 304) notes: 'I've climbed to the highest reaches of the planet in search of it.'

Sometimes the landscape is described as a type of dreamscape, where colours blur and merge, strange sounds are heard, and the traveller is transported into another realm. Mary Kingsley, normally all no-nonsense practicality, occasionally lets her fantasies take control, especially when she is by herself and at one with nature:

> I shall never forget one moonlight night I spent in a mangrove swamp. I was not lost but we had gone away into the swamp from the main river ... We got well in, on to a long pool or lagoon; and dozed off and woke, and saw the same scene around us twenty times in the night, which thereby grew into an aeon, until I dreamily felt that I had some-how got into a world that was all like this, and always had been, and was always going to be so. (Kingsley, 1897: 10)

She is also capable of the most poetic paeans to nature, referring to a palette of colours in words as vivid as any artist might employ. The moon is described as 'a great orb of crimson, spreading down the oil-like, still river; a streak of blood-red reflection', while the sun 'sank down into the mist, a vaster orb of crimson, and when he had gone out of view, sent up flushes of amethyst, gold, carmine and serpent-green, before he left the moon in undisputed possession of the black-purple sky' (Kingsley, 1897: 26). Kingsley also uses musical metaphors, with the Ogowé River 'as full of life and beauty and passion as any symphony Beethoven ever wrote: the parts changing, interweaving, and returning' (Kingsley, 1897: 27). This is a syn-ergy here with the views of polar trekker Liv Arneson, who argues that 'an expedition is a work of art expressed on a canvas of snow, air and time' (Arneson & Bancroft, 2003: 20–21).

John Muir's writing similarly has a strong spiritual quality, particularly when he is describing the Sierra in *The Mountains of California* (1894). It is 'so luminous, it seems to be not clothed with light but wholly composed of it, like the wall of some celestial city' (Muir, 1894: 295). Muir's choice of tran-scendent language is beatific and perhaps reflects his upbringing as the son of a Presbyterian preacher. He constantly refers to the landscape around him as 'sublime', 'glorious' and full of 'grandeur' and describes the natural world as something which thrills him to the very core:

> And after ten years spent in the heart of it, rejoicing and wondering, bathing in the glorious floods of light, seeing the sunbursts of morning among the icy peaks, the noonday radiance on the trees and rocks and snow, the flush of the alpenglow and a thousand dashing waterfalls with their marvellous abundance of irised spray, it still seems to me above all others the Range of Light, the most divinely beautiful of all the moun-tain chains I have ever seen. (Muir, 1894: 295–296)

These intensely spiritual experiences can lead to self-discovery and personal growth. David Breashears (1999: 242) wanted to share with others, through his film-making, 'the mountain experience ... the serenity of chasing away the darkness ... the transcendence of the mountain and the hope and awe it inspires'. He believed 'in exploring the terrain of that mountain, I was really exploring a far more mysterious terrain – the landscape of the soul'. This is the ultimate journey – the search for self-discovery or an *Absolute Other* (MacCannell, 1976) – which underpins much of our wanderings and the reluctance of many travellers to head home.

5 The Return

The final stage of the hero's journey, *the return*, takes place once 'the hero-quest has been accomplished' (Campbell, 1949: 193). This bestows mythic status, in that 'to look death in the face and to return to the living is the ultimate proof of a hero's extraordinary stature' (Van Nortwick, 1992: 28). The explorer traveller might be on a high, flushed with success, and ready to tell all, in the form of a book, documentary or lecture series – although increasingly the blog has told all throughout. They need time to rest, reinvigorate themselves and generally plan the next trip. Kira Salak recognised the pull of travel, but also the need to come home:

> Travel itself will always seem suspect to me; it is, after all, one of the most obvious forms of escapism. There is some other, better reality that we think we need, so we travel to find it. And maybe we do find it, or we don't, but the endless searches continue so that entire economies are fuelled by our insatiable need for Something More ... But of course we must always return. (Salak, 2001: 371–372)

Simply returning may be success enough for some travellers. For Jon Bowermaster, kayakking through the Aleutian Islands was seen as 'an incredible confidence booster. If I could pull this off – if we could pull this off – it would allow me, I envisioned, to undertake a variety of future expeditions, whether more physically challenging, logistically difficult, or simply more expensive' (Bowermaster, 2000: 62). After his expedition, he noted: 'For me, the confidence gained in coming to such a remote place and living in it, trying to understand it, and returning from it is plenty of success' (Bowermaster, 2000: 248).

However, not all goals are likely to have been met, which might lead to a sense of anti-climax or disappointment upon returning home. Cronin (2000: 65) observed in some instances that 'the promise of closure, of synthesis in return, becomes an open or festering wound'. Tragedy may have occurred, which blighted, for example, the homecoming of the *Terra Nova* after Scott's death in Antarctica. The survivors may feel tremendous guilt at returning home without their fallen comrades. For some, like Apsley Cherry-Garrard on Scott's expedition, this feeling never leaves them and overshadows the rest of their life.

Regardless of the outcome of the journey, explorer travellers may simply not want to return home. John Muir (1894) describes the mountains through which he has travelled as a holy place, but also where freedom can be found. He warns that 'few places in this world are more dangerous than home. Fear not, therefore, to try the mountain passes. They will kill care, save you from deadly apathy, set you free, and call forth every faculty into vigorous, enthusiastic action. Even the sick should try these so-called dangerous passes, because for every unfortunate they kill, they cure a thousand' (Muir, 1894: 328). Muir can only continue his quest to save the Sierras by continuing to travel. Others return but are desperately unhappy and restless. As Jack told us in an interview: 'I was getting addicted to travelling ... I was not happy at home any more'.

This chapter considers both the exhilaration and the difficulties inherent in the return home, including problems settling down and reintegrating into society. The traveller might be given great rewards or honours from their exertions, but some will struggle with everyday life after the exhilaration and pleasures of pitting themselves against the vagaries of a risky existence. Others seek a purpose or meaning in what they have done through inspiring others or highlighting a worthy cause. We also consider how the explorer traveller documents their journey and disseminates their accounts to the wider world, inspiring a new generation of travellers in an endless and circular process.

The Long Wait

The return of the explorer traveller is not assured, and may be prefaced by a long and anxious wait for loved ones. When should they give up hope? The Homeric myth of Penelope and Odysseus comes to mind, with Penelope waiting devotedly for 20 years for her husband to return from the Trojan Wars. There are modern parallels with the likes of Lady Jane Franklin and Nina Fawcett. Like Penelope, Lady Jane never lost faith, but did more than just endure her husband's absence. She organised various expeditions to find Sir John Franklin, who had disappeared after abandoning his ice-bound ship, during an attempt to find a way through the Northwest Passage. Leed (1995: 198) refers to *journeys of recovery* to find lost explorers as symbolic of the paradox of exploration – 'that so much of the new is discovered in attempts to recover someone or something perceived to be lost'. Lady Jane was notorious for her refusal to believe that her husband was never to return home. She funded seven expeditions from her own fortune and supported many more, to search for Sir John. Only her death in 1875 stopped this mania in its tracks. Even when Lady Jane realised he wouldn't return, she sought his papers, to 'learn whether he had traversed the undiscovered passage before he died' (Leed, 1995: 203). She therefore had an eye to posterity, and wanted credit for his achievements, however much it cost and however long it took. Nina Fawcett also obsessed over her husband Percy's return from the Amazon after his disappearance in 1925. She combed through records, followed up leads, and was never at peace, wandering the world like a lost soul (Grann, 2009). Nina died still believing her husband and elder son were alive. Other searches for lost explorers include Stanley's quest to find Livingstone, the rescue expedition for Burke and Wills and the search for Mungo Park (Leed, 1995). Park returned home in 1804 after being lost for 10 years, but subsequently disappeared again in Africa while on a second expedition and was never found. In fiction, the search for the lost explorer is a common trope (see Chapters 6 and 7).

The flipside of waiting is the fear of being forgotten by loved ones, or that things have changed between them. Leed (1995: 223) notes that: 'Heinrich Barth was crushed when, after three years of exploring and meticulously describing the country along the Niger to Timbuktu, he returned to find his employers had accepted – too quickly, it seemed to him – the rumour of his death as fact and already sent in his replacement'. Douglas Mawson writes poignantly to his fiancée Paquita on 26 December 1913, after receiving her letter written the previous November, in which she included the sentence 'Dearie, I hope you and I are going to be happy':

> My Darling, it lies with you – *Can you be happy with me*. I have aged in appearance with this strain and may not appeal to you now. My body tissues have been strained and cannot be so good. But at heart I am just the same though perhaps more impressed with your qualities. Size me up critically, and *don't* let us get married *unless* after reflection you feel nothing but attraction and an abandonment to my desire just as I feel to yours. (quoted in Robinson Flannery, 2000: 125)

He had nothing to fear. They were married, it would appear most contentedly, for 44 years, until his death in 1958. Mawson kept her letters for the rest of his life (Robinson Flannery, 2000).

Recrossing the Threshold

Like the departure, the return of the traveller involves stepping over the threshold, an act which symbolises *home*. It might be an airport, harbour or front door. A number finish back where they started – at the Royal Geographical Society. Tim Cope finished his 2005 trek across Kazakhstan with a tour of the RGS headquarters and a formal lunch there the next day. There is usually some ceremony attached to the return, marking that they have gained social distinction as a result of their exploits (Kane, 2010).

Formal awards and public acclamation may follow. The RGS awarded gold medals to a select few, including Livingstone and Burton (Murray, 2008). A number received knighthoods, including Sir Richard Burton, Sir Ernest Shackleton and Sir Ranulph Fiennes, and became household names when they returned. Few in 1953 could have been unaware of Sir Edmund Hillary, knighted immediately after his successful climb of Mt Everest, even if he was one of the most retiring and modest of adventurers. Images of explorers and adventure travellers graced official currency and stamps. Australian examples include James Cook, Charles Sturt, Hamilton Hume, Matthew Flinders, George Bass and Douglas Mawson. Statues and memorials are another legacy of feats of exploration. Statues of Robert Falcon Scott can be found in Christchurch, New Zealand and Portsmouth Historic Naval Dockyard, while Livingstone's statue graces Edinburgh, Glasgow and Victoria Falls. A memorial to Burke and Wills stands at the corner of two of the busiest streets in Melbourne, showing the explorers in a larger than life heroic pose, symbolically returned in triumph (Figure 5.1). The French government laid a plaque in Timbuktu in 1903, given by the Council of the African Society to honour Alexander Laing's achievement in reaching the fabled town, 77 years after his murder (African Affairs, 1964).

Richard Burton's celebrity was a joint effort – partly created by his own larger-than-life persona as depicted in the pages of his writings, and continued by his wife after he died. He cared about his image, asking his portraitist Leighton not to 'make me ugly' (Murray, 2008: 46). His wife edited his writings after his death, to ensure that they contributed to rather than detracted from his heroism, and built him a tomb in the shape of an 'Eastern traveller's desert tent' (Murray, 2008: 44), lest anyone should be tempted to forget why he was famous.

Roald Amundsen found that his return was more heralded than his departure. He was quietly pleased at the attention his arrival on the *Fram* in Buenos Aires received after news spread of his feat in reaching the South Pole: 'At our departure there were exactly seven people on board to say

Figure 5.1 Burke and Wills Memorial, Melbourne
Source: J. Laing.

good-bye, but as far as I could see, there were more than this when we arrived; and I was able to make out, from newspapers and other sources, that in the course of a couple of months the third *Fram* Expedition had grown considerably in popularity' (Amundsen, 1912: 637–638).

Not all adventurers, however, are afforded public acclaim for their efforts. Yet even if this fame is just within their own social circle, it allows the explorer traveller to tell a compelling story at a dinner party (Adler, 1989) and to impress family and friends. Polar trekker Ann Daniels wanted to make her children proud (Hamilton, 2000), while Michael told us he sought his father's approval for his adventure pursuits. The return might also be an opportunity to thank those close to the traveller for their assistance with the journey. Patrick Woodhead, on his return from Antarctica, simply wanted to tell his parents 'how much I had valued their support. How their affection had been one of the only things keeping me moving on that last desperate night to the pole' (Woodhead, 2003: 321). Matt Dickinson, after returning home from climbing Mt Everest, was asked by friends whether he was going to be invited to Buckingham Palace or the Royal Geographical Society. His reaction to this is pure British understatement but perhaps his extreme humility is a little hard to believe: 'In fact nothing happens at all, which I find extremely refreshing. The nearest I got to a fanfare was a poster my son Alistair painted in crayon daubed with the legend "Ruddy Well Done Dad!" That was tribute enough' (Dickinson, 1998: 209).

It is more common, however, for explorer travellers to find this lack of attention disconcerting. Henry Morton Stanley was mortified after he returned from finding Livingstone in Africa to receive a less than effusive welcome. The RGS released a statement that made it clear that they could not 'hold functions in the summer, and so could not entertain Mr Stanley' (Jeal, 2007: 139). Their reluctance to fête his achievements was perhaps due to his success at their expense – the RGS expedition to find Livingstone had been a failure – as well as the sheer implausibility of a mere journalist becoming an explorer, which led some newspapers to speculate that he might have fabricated his story (Jeal, 2007).

Many explorer travellers were quite up-front about their desire to be famous. The sailor Hugo Vihlen explained, 'the trip has been a great experience and a test of myself in many ways, but it's not something that I would do if I thought no one else in the world cared about it' (Vihlen, 1971: 183). Balloonist Colin Prescot was also honest about the attraction of the limelight: 'to a certain extent the anticipation of an element of glory also adds to the attraction of the kind of derring-do in which I have indulged' (Prescot, 2000: 16). Prescot saw a parallel to Richard Branson in terms of their motivations:

I suppose you could say we both had a similar, rather childlike obsession with adventure. More specifically, we shared a passion for the eccentricity and theatre of adventure. This could be summed up as showing off. In other words, I suspect the two of us would agree that none of our capers would be nearly as much fun if no one was watching. (Prescot, 2000: 137)

While long-distance balloonist Bertrand Piccard disclaimed any interest in becoming a celebrity ('We flew for the passion of flying, of exploring all the skies of our planet'), when he landed in Switzerland he was disappointed at the reception, observing: 'what a pity we didn't get back yesterday, at the weekend, because a lot more people would have been free to come and meet us' (Piccard & Jones, 1999: 285). His partner Brian Jones was even more direct about his displeasure at the lack of a crowd: 'they could have given one of the cleaners a Swiss flag ... and she could have come out on the balcony and waved it.'

The return brings with it a mix of emotions. What now? What next? Laurens van der Post (1958: 253) wrote of driving 'back to our twentieth-century world' after travelling through the Kalahari desert. He used his experiences as the basis of books and a documentary film, which shone a light on the Bushmen of Africa. Others did not know whether they would now fit in at home after their extraordinary experiences. Cronin (2000: 65) argued that the traveller 'who has been to foreign parts is not only *unsettled* but s/he becomes on return an *unsettling* figure for the settled community'. Tim Cope, after his first expedition through China and Russia by bicycle in 1999, faced uncertainty after being on the road for so long:

> It struck me that I didn't have a clue about my future. Beyond this journey was just a big unknown blank. I did not have the slightest idea what I would do when I returned to Australia. I wasn't going to study and I would probably have a big debt. What did Australia mean? Surely I would feel like a fish out of water back there. It was here in the northern forest that I felt most alive. (Cope & Hatherly, 2003: 32)

Others see travel as a continuum – the gaps between journeys are effectively non-existent – and thus they don't experience the return as a low point or reason for gloom, but merely the start of the next adventure. As Sara Wheeler writes at the end of her journey across the small Greek island of Evia: 'I felt curiously dispassionate about my imminent departure. Nothing seemed to have ended and nothing would end' (Wheeler, 1992: 279). She quotes Laurens van der Post: 'Where the body stops travelling, the spirit takes over the trek'.

The Restless Return

There are numerous accounts of the disappointment suffered by explorer travellers when they return, of the let-down of coming home, and difficulties in settling down. Paquita, the then fiancée and later wife of Douglas Mawson, alluded to this in one of her letters to Mawson while he was in Antarctica:

> I don't want to doubt you my dear but I'm afraid of the fascination of the South. All the members say they would go again & here is Shackleton off again. Will a calm life ever satisfy you? I have seen unhappiness where I thought all was well. Calm homes also have skeletons in a cupboard it seems. I want you to reassure me that all will go well with us and our love. (quoted in Robinson Flannery, 2000: 102)

While he largely built an academic career upon his return, Mawson returned to Antarctica for the 1929–1931 British Australian and New Zealand Antarctic Research Expedition, and thus arguably did not keep his promise to Paquita 'that I shall never cause you such trouble again' (Robinson Flannery, 2000: 116).

Others have a similar story to tell of the importance of travel in their lives. Mountaineer Andrew Lindblade could not overcome his restlessness, yet recognised the hazards of heeding the siren call:

> My times in the mountains so far have been very intense and at times violent, yet it is in these moments – despite being unaware of it at the time – that I have felt most contented. After returning home I leave the 'presence' of a completed ascent and attach myself to the potential of another. Then there comes an unmistakable time: I feel something in the wind. The peace of home soon subsides, and I champ at the bit to get back into the mountains, back into the thick of things, back where peace and total chaos live side by side. But then, the predatory, paradoxical question surfaces: surely I could find peace and contentment without having to reach out and up into the dangerous mountain world? (Lindblade, 2001: 2)

Others ask a similar question, yet acknowledge that nothing else will make them feel the same or bring them the same kind of peace. Climber Anatoli Boukreev felt 'melancholy ... in my ordinary life' which he hoped 'will pass when there is another magnificent peak'. He wondered, however, whether 'this longing and restlessness [is] the price that mortals pay for daring to trespass in the houses of the Gods ... the price you pay for disturbing the peace of God?' (Boukreev, 2001: 36).

Separated from travelling and life on the road, individuals can experience almost a kind of grief at not having somewhere else to go to. Jeffrey Tayler, at the end of his Saharan desert expeditions, describes how he 'felt a painful sense of bewilderment: for the first time since Tizgui, I had nowhere left to go' (Tayler, 2003: 237). The solution is often to plan another trip. Arriving in the wintry chill of a Liverpool January, Mary Kingsley (1897: 15–16) cast her eyes back towards Africa: 'if you do fall under its spell, it takes all the colour out of other kinds of living'. Going back, for her, is inevitable. Similarly, even after his successful Everest climb, Dickinson (1998: 210) still 'can't last more than a few days at home before I am pacing the floor thinking about where I am going next. I still lie awake in the early hours of the morning tracing journeys in my mind through the places I haven't yet been'.

Transformation

Not all returns are disappointing. Many revel in how travel has changed them. The transformative potential of travel is a key motif in many books – that seductive promise that we will never be the same again after a journey (Laing & Frost, 2012). As Jon Muir (2003: 67) wrote: 'What great lessons will the Great Mission teach me? How will it change me?' The likelihood of further transformation can be a reason for returning to the road. Charlie alluded to this in his interview: 'I think I'd approach [another solo trip] in a different way. I'd be much more honest with myself in a way … yeah, it would be [a] far more exciting me. I think I'd just be more broad-minded'.

Travel to certain frontier places is particularly linked with transformation. Patrick Woodhead is fascinated at how travel through Antarctica, for example, changes people:

> In almost every account I have read, there is a recurrent theme; people claim to have been in some way 'touched' by their experiences on the ice. Some even go farther and describe how, when they returned, they found it hard to integrate themselves back into normal life. Antarctica, it seems, had had some sort of life-affirming effect on them. (Woodhead, 2003: 104)

There may be a sense that this is a self-fulfilling prophecy – something that is expected and perhaps manufactured, to a degree. Ben Kozel, at the end of his journey, came back 'changed' as he expected to: 'I had gone to South America expecting to come back feeling changed in some way, able to apply some new attitude in the way I went about my day-to-day life' (Kozel, 2002: 319).

All the examples discussed above involve men. Women provide similar narratives about being transformed by explorer travel, but this occurs against a very different backdrop, where the mere fact of travel is often a cause for celebration and a 'platform for transformation [where] women's business and pleasure travel experiences create opportunities for them to gain empowerment in themselves; power that is also reflected to others' (Harris & Wilson, 2007: 239–240).

Feminine emancipation

Women's experiences as explorer travellers are less well understood, perhaps because historically there have been fewer of them and their narratives might not have received such wide circulation or approbation. Blunt (1994: 32) argues that women are called travellers, 'but rarely explorers, suggesting constructions of the overt masculinity of exploration and the more passive femininity of travel'. Even the term traveller is more likely to bring with it masculine overtones, with a woman's place more readily associated with home and hearth (Smith, 2001). The stories of female adventurers are often absent from libraries or the public imagination. Liv Arneson was aware of this gap in gender narratives even as a child: 'Weren't there any adventurous women to read about?' (Arneson & Bancroft, 2003: 17–18). Kira Salak noted that:

> the lone woman traveller is still seen as such an anomaly. When are little girls taught to fear camping by themselves? To fear wild animals or insects or mud? Or travelling on their own? They must be *taught* all of this, at some point, but I don't know when. (Salak, 2001: 378)

As always, there are exceptions. Female explorers like Mary Kingsley and Isabella Bird were high-profile travellers in their day, whose travel accounts remain popular and in print (Whybrow, 2003). They travelled in Victorian times, when a lady was expected to live a largely sedentary and decorous life. Escaping the confines of their restricted lives back home was often the making of them, improving their health and giving them a reason to live. This is ironic in the case of Mary Kingsley, who was honest about the dangers of her journey to Africa and wrote that she went out there to die (Wheeler, 2007). In a sense she did, having been freed from the burden of caring for her parents, which left a void that travel ending up filling.

This feminine desire to engage in adventurous travel is not just a historical phenomenon. Caroline Hamilton comments on her 1997 polar trek: 'Mothers, students, lawyers, business women, teachers and artists all wanted to be part of [my] expedition. It seemed that everywhere, in all walks of life, women were looking for something more, something different – a challenge to push them to the very limits of themselves' (Hamilton, 2000: 16). Her teammate, Rosie Stancer, describes in vivid terms how her dreams came alive: 'It was as if a lit match had been dropped on to paraffin' (Hamilton, 2000: 17), while Kira Salak's journey to Papua New Guinea is 'bred from years of desperation' (Salak, 2001: 371). For Salak, travelling solo, with its attendant risks, is about transforming herself, returning as someone new, 'with an almost superhuman strength and confidence' (Salak, 2001: 370). This desire for adventure may be linked to the freedom which comes from leaving behind the rules, restrictions and societal conventions which often constrain feminine leisure pursuits, linked to women being the primary caregivers and 'the gender role that is typically expected of them' (McGehee *et al.*, 1996: 47–48).

For these women, proving they are as good as men involves undertaking the same sorts of hardships, without complaint. Mary Kingsley had 'behind me the prestige of a set of white men to whom for the native to say, "You shall not do such and such a thing", "You shall not go to such and such a place," would mean that those things would be done' (Kingsley, 1897: 102). Her constant reference to herself as a man is argued to illustrate her ambiguity about being a woman in a man's world, and to represent an attempt to position herself 'within the masculine tradition of scientific observation' (Blunt, 1994: 76). Kingsley therefore sought to emulate these 'fine specimens', although 'it was hard to live up to these ideals' (Kingsley, 1897: 102–103). She describes hellish conditions and the fatalism of encountering disease as a form of Russian roulette: 'no man knows from day to day whether he or those around him will not, before a few hours are over, be in the grip of malarial fever, on his way to the grave' (Kingsley, 1897: 260). This journey to Africa resulted in a 'metamorphosis ... It was impossible now to slip back into the domestic servitude, the feminine self-immolation of her pre-African life' (Frank, 1986: 90). Some have voiced disappointment that Mary Kingsley did not translate these bold deeds into support for women's rights back home. Unlike Isabella Bird, she did not lobby for women to be able to join the Royal Geographical Society, nor did she support women's suffrage (Blunt, 1994; Frank, 1986). Nevertheless, the mere fact of her adventurous journeys and the accounts she wrote of them inspired other females to travel and showed that it was possible to transcend societal expectations of what was suitable behaviour for a woman.

Isabella Bird was similarly stoic. Her first book, *An Englishwoman in America* (1856), was anonymous, but thereafter she published under her own name. *A Lady's Life in the Rocky Mountains* (1879) tells the story of her often solo journey through the Rockies, although she describes with great honesty, and sometimes a disparaging tone, the people she meets on the way. She provides lyrical and highly emotional accounts of the beauty of the landscape, describing Lake Tahoe as 'a dream of beauty at which one might look all one's life and sigh' (Bird, 1879: 1) and the views on either side of the Truckee River, where 'great sierras rose like walls, castellated, embattled, rifted, skirted and crowned with pines of enormous size, the walls now and then breaking apart to show some snow-slashed peak rising into a heaven of intense, unclouded, sunny blue' (Bird, 1879: 11). Bird takes her carpetbag to some of the most iconic parts of the American West, and faces tremendous dangers, which she acknowledges at times, suffering 'fatigue, giddiness and pain' (Bird, 1879: 110) when she ascends Long's Peak, and being thrown off horses and surviving rattlesnakes. At other times she plays down the risks, rather like Robyn Davidson in *Tracks* (1980), as if conscious of appearing too self-congratulatory (Holland & Huggan, 2000).

Davidson is however not coy about her strong sense of achievement at the end of her trip, despite the fact that she finds her 'new adventuress's identity kit ill-fitting and uncomfortable' (1980: 252). She writes that the two important things that she learnt from her trek across the Outback are: 'that you are as powerful and strong as you allow yourself to be, and that the most difficult part of any endeavour is taking the first step, making the first decision' (p. 254). Yet she also admits her disappointment that others may not acknowledge or understand her personal growth or the challenge inherent in her travels, with her observation about 'defending myself against people who said things like, "Well, honey, what's next, skateboards across the Andes?"' (p. 252). Thus her return has shown her that women adventure travellers are often treated as curiosities, rather than the equal of their male peers.

Constructing the Narrative

Jean-Paul Sartre wrote: 'For the most commonplace event to become an adventure, you must – and this is all that is necessary – start recounting it ... But you have to choose: to live or to recount' (Sartre, 1938: 61). Tales of exploration have a long history and may shape and inspire travel experiences by setting 'the imaginative pathways for contemporary adventure tourists' (Zurick, 1995: 49). Many explorer travellers have written books or journals about their travels, informing and motivating people with their accounts of the exotic and dangerous settings in which they found themselves, and providing a historical record of their motivations for undertaking their journeys. These texts, and the type of travel they are based on, can be labelled *metempsychotic*, where 'the tourist takes on the persona of a significant other or group, as a role model for a particular repeated journey' (Seaton, 2002: 155).

There are many references to keeping diaries and notes, with the aim of publishing this when one returns. William Dalrymple's *In Xanadu* (1989) regularly mentions his diary or 'logbook' entries, written in exotic locations during his re-enactment of Marco Polo's journey to visit Kublai Khan. The book itself might not be written until one is home, or on the return journey. Amundsen wrote his account of his expedition to Antarctica in Brisbane:

in the shade of palms, surrounded by the most wonderful vegetation, enjoying the most magnificent fruits, and writing – the history of the South Pole. What an infinite distance seems to separate that region from these surroundings! And yet it is only four months since my gallant comrades and I reached the coveted spot. (Amundsen, 1912: 25)

Syndication rights are like gold to an impecunious adventurer. Fawcett sold the rights of his account of an Amazonian journey in 1924 to the newspapers. Brokered by a war correspondent he met on the road, the story was bought by a North American consortium and then later by newspapers across the globe (Grann, 2009). This trained a spotlight on Fawcett's expeditions, which led to much-needed funds and support from scientists and scientific organisations who also subscribed to Fawcett's theories of a lost world ready to be discovered. One was the American Geographical Society, whose director had been on Hiram Bingham's 1911 expedition that found Machu Picchu (Grann, 2009).

Controlling the story could, however, be controversial. When the *Terra Nova* returned from Antarctica after the tragic conclusion to Scott's expedition, it anchored off the coast of Oamaru. Several men rowed ashore on 10 February 1913 and asked to send a cable to their New Zealand agent with news about the expedition. They refused to provide any details and the cable was sent in code. Rumours spread that one of the men was Scott himself and officials were even falsely misled that Scott was on board. The local newspapers were denied the scoop of the century, in order that the expedition's magazine sponsors were the first to find out what had happened to Scott and four of his expeditioners (Bruce, 2013). The centenary of this turn of events was commemorated in Oamaru in February 2013 (Figure 5.2). Ironically, Amundsen, Scott's rival, had tried to keep his 1912 journey to the Pole secret, not to protect syndication or sponsors' rights, but to stop his 'project being stifled at birth' (Amundsen, 1912: 59). He thought reaching the South Pole first would allow him to raise funds for his original plan to explore the North Polar basin. On their way to Antarctica, the *Fram* anchored in Funchal, and Amundsen was dismayed that the newspapers assumed he was heading for the South Pole and 'had no idea of the value of the startling piece of news they were circulating'. Luckily for them, the news 'did not fly beyond the shores of Madeira' (Amundsen, 1912: 129) and their secret remained safe for the time being. In these days of social media, it is hard to believe that such secrecy could be maintained.

Some explorer travellers emphasise the impact of the books they write and this provides a pro-social justification for their exploits, an outcome of adventurous journeys discussed in more detail later in this chapter. For Graham, writing about his experience allows him to share it with others and gives it an altruistic flavour: 'I've written either a book or a magazine article about all the expeditions I've done. And that just makes it for the broader community, rather than just doing something for yourself'. Similarly, Liv Arneson has published a book in Norwegian titled *Nice Girls Don't Ski to the South Pole* and is candid about its motivational impact on her female readership:

> The response to the book in Norway was incredible. I kept getting letters from women who told me that reading about what I had accomplished

Figure 5.2 Re-enactment of the sending of the coded telegram, Oamaru Scott 100 Celebrations, February 2013
Source: J. Laing.

when I was forty-one made them believe in new possibilities for themselves ... I began to see the powerful effect that my experiences were having on other people. And that connection was as important and profound to me as the trips themselves. (Arneson & Bancroft, 2003: 28)

These books make an explorer's reputation assured and can help to fund further expeditions. David Livingstone earned £8500 from *Missionary Travels and Researches in South Africa* (1857), in comparison to the £100 he received per year as a missionary (Murray, 2008). Some texts became bestsellers due to the momentous events they capture, which the adventurer could not predict before they left home. Jon Krakauer was not to know that the Everest expedition he joined as a journalist would result in numerous deaths and help to make *Into Thin Air* such a compelling and tragic read. The book received many awards, including *Time* magazine's Book of the Year, and is justly regarded as a classic.

A variety of media is used to share adventurous experiences with others. Peter, for example, used documentaries 'to record and show people all the great things that I was doing because I was so blown away when I first did it and you want to be able to tell people about it and show them what it's like'. Aaron did this through his website: '[The internet] was a great way to share round the world with hundreds of thousands, maybe millions. I don't know how many people became exposed to this kind of thing'. Hugh Thomson, on his return from Peru, wrote a book *The White Rock: An Exploration of the Inca Heartland* (2001), and co-curated an exhibition on Hiram Bingham and his discovery of Machu Picchu for the British Museum and Oxford and Cambridge universities, as an way of championing the explorer and facilitating greater recognition of his achievements.

Lecture tours and speaking engagements are another way of making the return meaningful and giving the traveller a purpose back home. Mary Kingsley was popular with audiences and spoke around the country (Frank, 1986) but made little from her public lectures, being 'subject to the lower lecture fees and limited venues for women generally' (Blunt, 1994: 72). These lectures, however, helped her writing by giving her feedback on her musings, and generating interest in buying her books, as well as giving this socially awkward woman a vehicle with which to reach out to others (Frank, 1986).

A number of people we interviewed referred to the inspiration they gained from attending public lectures. Helen remembered the impact of a talk by a Sherpa at her school: 'I just knew straight away, the thing I wanted to do was go and see these places, like the Himalaya. I wasn't sure, but I just wanted to eliminate that query that was big in my mind since seeing those pictures and I'd read all these books. I was just about to explode with excitement!' Doug as an adult saw a slideshow given by two climbers who had climbed Everest: 'I saw it, was swept away by it and thought I'd like to do it'. The presenters are not unaware of this power they wield to deliver persuasive messages through their public lectures. Mountaineer Conrad Anker wrote that he uses slide shows as a way to share his Buddhist outlook on life:

> People come to see slides of me climbing, to share my adventures, but I can use the opportunity to talk about being a good person, about how anger and hatred disrupt an expedition, about how sometimes it takes a little more effort to be positive than negative, but that's ultimately life-enriching. I'd like to take what notoriety or fame comes my way and turn it into something good, as for instance Sir Edmund Hillary has, building schools and hospitals in Nepal. I'd like to share what mountains have done to change my life, and become a spokesperson for goodness. (Anker & Roberts, 1999: 102)

Yet not all accounts can be trusted and examples of yarns or exaggeration are numerous. *The Travels of Marco Polo* appears to have combined fact and fiction (Laing & Frost, 2012) and some argue the entire work is a fake (Wood, 1996), although the counter argument is that the level of detail makes it difficult to suppose that it was a mere invention (Nevett, 2004). Other disputed journeys include the claims by Peary and Cook that they reached the North Pole (Conefrey & Jordan, 1998) and Marlo Morgan's *Mutant Message Down Under* (1991), discussed in Chapter 9. This suspicion over the veracity of claims of exploration reached its height with the conspiracy theories that surround the Moon landings. Hollywood took this idea and turned it into the film, *Capricorn One* (1978), which focused on a faked Mars landing. When their spacecraft is burnt up upon entry due to a fault, the astronauts realise that they will be killed to ensure the public ruse is maintained, unless they escape. The conspiracy is finally exposed when one of the astronauts turns up at his own funeral, reminiscent of Twain's *The Adventures of Tom Sawyer* (1876). A number of documentaries cast doubt over whether human beings have ever reached the Moon and bluntly label the Apollo programme a fraud. This view persists in some quarters, despite the evident inherent difficulty of keeping such a secret from the thousands of employees who worked on the programme over the years. Even the television footage has been scorned as a clever fraud. It is extraordinary that such nonsense persists.

The Spoils of Exploration

The mania for mapping underpins much of exploration, based on the idea that we do not really discover a place until it is named and mapped. Mary Kingsley refers to the source of the Niger and how this *problem* was finally solved by explorers like Mungo Park, yet this conundrum and the loss of life it occasioned 'may seem very strange to us who now have been told the answer to the riddle; for the upper waters of this river were known of before Christ and spoken of by Herodotus, Pliny and Ptolemy ... but they were not recognised as belonging to the Niger' (Kingsley, 1897: 11). Consistent with a desire for organisation and order, these travellers 'transformed the profligate chaos of the entire globe into the orderly Latinate cosmos of Linnaean taxonomy' (Smith, 2001: 3).

Naming rights are one of the 'chief rewards of discovery' and could be 'traded for support and patronage'. It is all about being first, but also that this precedence be 'recognized by "the" public and its representatives' (Leed, 1995: 211–213). That there was often an indigenous population and a local name already in existence is usually overlooked. It took 120 years for Ayers

Rock, named by William Gosse in 1873 after the South Australian Premier, to have its original name of Uluru restored (Hodges, 2007).

Things are more straightforward in Antarctica, due to the absence of a local population. Unlike Australia's conquest, the *terra nullius* argument for annexation of land in Antarctica is less disingenuous and was based on the actions of *explorers* rather than settlers. The argument of 'land belonging to no one – can be established through the eyes, feet, codified ritual performances, and documents of explorers' (Collis, 2004: 50). Amundsen arrived at the South Pole and performed a ceremony: 'We all assembled about the Norwegian flag – a handsome silken flag – which we took and planted all together, and gave the immense plateau on which the Pole is situated the name of "King Haakon VII.'s Plateau"' (Amundsen, 1912: 15). Naming was a precursor to staking out territory. In 1930, 'Mawson affixes names to every landform he can see: the black island becomes Proclamation Island; inland mountains are assigned the names of the men ... A crucial moment in Antarctica's spatial history has occurred: Antarctica has been produced as Australian space' (Collis, 2004: 43).

Naming may be accompanied by marking the territory with an initial or signature. Livingstone discovered and named Victoria Falls in Africa after Queen Victoria, which he stated was 'the only English name I have affixed to any part of this country'. He also carved his initials and the date on a tree, again the only time he said he 'indulged this piece of vanity' (quoted in Murray, 2008: 309). Lewis and Clark were notorious for the practice. Both carved their names on a tree when they reached the Pacific Ocean. When the tree was destroyed, it was replaced by a commemorative sculpture for their bicentenary. Clark also carved his name into a sandstone butte on the Yellowstone River, which he called Pompey's Tower (now Pompey's Pillar), after Sacagawea's son. Sometimes there can be conflict over who has the right to bestow a name on a place. The Americans named a piece of land in the Canadian Artic *Grinnell Land*, but the British disputed this, arguing that it was part of land they had already discovered and called Devon Island. To avoid a diplomatic incident, it was called 'The Grinnell Peninsula of Devon Island' (Leed, 1995).

Causes and Charities

Explorer travellers are often keen to put something back into society after they have finished their adventurous journeys. These *pro-social* motivations, defined as based around 'service to others' (Anderson & Shaw, 1999: 103) are often seen in the volunteer tourism sector (see, for example,

Jago & Deery, 2001; Ooi & Laing, 2010; Ryan & Bates, 1995; Singh & Singh, 2004). In the context of adventure travel, it is often grounded in a desire not to appear self-centred and acknowledges their good fortune and privilege in undertaking these extraordinary journeys. Graeme Dingle, the New Zealand climber who set a number of records, including being the first person to trek the entire length of the Himalayas and to circumnavigate the Arctic Circle, notes that his adventures were 'selfish' and this motivated him to set up the Sir Edmund Hillary Outdoor Pursuits Centre to involve schoolchildren in outdoor education (Kane, 2010). Michael told us how he partly assuages his guilt with respect to his travel through raising money for charity:

> I've found myself working for the Women and Children's Hospital, the Westmead Kids' Hospital and the Smith Family, through their Learning for Life program, which is an educational program which gives money to families with no money; multi-generational poor families, to give kids an opportunity to try and get an education and then achieve their potential. And the parallel is, doing the trips is kind of achieving my potential. [There's] no guarantee I'm going to make it but I'm certainly going to give it a try ... With the trips, I give away all the money I make, over and above the very considerable cost of making the trips happen. It's a half a million-dollar exercise to go to the South Pole. And I wouldn't have it any other way because the trips are fairly self-indulgent things.

Others use their profile as a springboard to inspire others. Aaron told us that he saw his polar treks as a vehicle for changing people's lives: 'I felt like I had an opportunity and a responsibility to have a positive impact on others. And so, when I did this, I went with the premise that I'm not just doing it for myself but I'm also trying to [do] good for others, maybe expand their horizons a little, their minds.' Like Michael, he mentions the guilt associated with his indulgent travel: 'I guess I needed that just to justify going and spending all this money when I could have been helping others.' Many female explorer travellers told us how they enjoyed being a role model for other women. As Karen observed: 'I'd love to encourage other women to get out and have a go at some big exciting things'.

These journeys can use the media exposure to highlight a message or to educate others (Laing & Crouch, 2011). Richard is candid about the power of the publicity generated by his diving trips to push his environmental causes: 'Because we get media with the things we are doing, [we] utilise it ... and it's a mighty soapbox to stand on when you can actually produce a documentary and know it's going to appear around the world'. Many re-enactment journeys are carried out to further knowledge, either to correct a misconception or to resurrect the reputation of a fallen hero – the explorer who has been vilified by modern narratives or written out of or overlooked in modern history books. Geoff told us that the story of Sir John Ross had been forgotten and how he hoped his re-enactment would remedy that: 'the story of how they actually got their people out from that area after four winters in record temperatures of minus 72 ... it's got to be told'. Robert Swan saw his expedition in the footsteps of Scott as a way to 'set the record straight on the gallant man's behalf' (Mear & Swan, 1987: 17) and writes of his 'obsession' with history: 'What is the real story of Scott? I have a feeling that we may find it out there on the journey' (Mear & Swan, 1987: 84). We discuss re-enactments more fully in Chapter 8.

Pro-social motivations can enrich the experiences of explorer travellers, giving their journeys meaning and a sense of purpose. Mitchell observed to us: 'Just knowing that you're making a difference to people's [lives], no matter how, even if it's just a tiny difference just to one person, it's worth it. So that, for me, was a big source of motivation'. Another example is Ann Bancroft, who talks of her early days as an expeditioner, and describes her ambition as ultimately an empty motivator, compared to the joy she found in inspiring others:

> It wasn't until I returned to the school where I had taught and saw how excited the kids and teachers were to have shared this adventure with me that I understood. The emptiness came from not having a full purpose beyond my own ambition. That's when I swore that if I ever did another expedition, I would figure out a way to bring the kids with me. (Arneson & Bancroft, 2003: 19–20)

Failure to Return

The return of the traveller may be as hazardous as their departure, when the mind is fixed on home, and weariness or despondency sets in as the adrenaline rush of reaching a goal subsides. The climber David Breashears notes: 'Remember that getting to the summit is the easy part; it's getting back down that's hard' (quoted in Krakauer, 1997: 290). All the deaths on Everest recounted in Krakauer's *Into Thin Air* occur on the descent, some close to camp (see Chapter 10).

It is relatively straightforward but sobering to reel off examples of explorers who did not return home. Mungo Park and Alexander Gordon Laing perished in 1806 and 1826, respectively, while trying to discover the source of the Niger in Africa and, in Laing's case, the fabled town of Timbuktu (Allen, 2002; Frank, 1986; Kryza, 2006). Park was drowned trying to escape the attack of an indigenous tribe, while Laing was murdered after leaving Timbuktu. Only a few years after Stanley found him in central Africa, David Livingstone died 'emaciated and wasted ... still in quest of the source of the Congo' (Frank, 1986: 30–31). Burke and Wills did fulfil their goal of crossing Australia from south to north (Melbourne to the Gulf of Carpentaria), but missed connecting with their base camp at Cooper's Creek. Following a punishing forced march they reached their base to find it had been abandoned earlier that day. They died soon after (Allen, 2002; Murgatroyd, 2002). Mallory and Irvine died in 1924 on an ascent of Everest and knew the fearful odds they took. According to Anker and Roberts (1999: 92), Mallory told his friend Geoffrey Keynes before his last expedition to Everest in the spring of 1924: 'This is going to be more like war than mountaineering. I don't expect to come back'. Davis (2011: 573) argues that Mallory's generation of climbers, all Great War veterans, took great risks in the mountains because they had cheated death before: 'They had seen so much of death that life mattered less than the moments of being alive'.

Some of this tragic news took months and even years to percolate through to those back home. The delay in finding out about Scott's death can be illustrated by the introduction to Amundsen's book by Fridtjof Nansen, written on 3 May 1912, in which he states: 'When in a year's time we have Captain Scott back safe and sound with his discoveries and observations on the other route, Amundsen's results will greatly increase in value, since the conditions will then be illuminated from two sides' (Amundsen, 1912: 21). Scott had already died by the time this was written, news that only became public in February 1913, after the *Terra Nova* reached Oamaru in New Zealand and telegraphed ahead to London.

Some did not perish on the journey but found the return difficult to endure. Apsley Cherry-Garrard suffered several nervous breakdowns over the course of his life. It was believed he was haunted by the guilt of surviving a trip to Antarctica while his comrades on the Winter Journey perished with Scott on their return from the South Pole (Wheeler, 2003). Others were believed to have taken their own lives. Speke, on his return from discovering the source of the Nile, was greeted by the news that his achievement was disputed by his rival Burton. He was challenged to a debate at the Royal Geographical Society. The day before the debate, Speke shot himself while game-shooting. While it was dubbed accidental, rumours of suicide were bandied about (Murray, 2008), notably by Burton, who claimed that Speke was in fear of the debate's outcome. While Jeal (2011) suggests that suicide was unlikely, this interpretation of events persists. Sadly, Speke's claims about Lake Victoria were posthumously found to be correct. Another who took his own life was Meriwether Lewis, three years after returning from his expedition. His depression was compounded by debts and an inability to bring his journals together for publication (Ambrose, 1996).

It has been argued that we tend to remember the fallen heroes, not those who made it back in one piece (Kelly, 2003; Laing & Frost, 2012). This is a high price to pay for everlasting glory. Yet many were happy to pay it, and still go on doing so. Their motivations for adventuring are complex, but fame and fortune are not the whole story. By and large, they believe that what they do has merit in bringing out the very best that human beings are capable of and see travel to the unknown as our destiny – something we are born to do. Thus their lives are willingly sacrificed in the pursuit of this higher purpose. There are two individuals in particular whose failure to return we wish to honour in this book. They are two of our interviewees who passed away while on adventures – Helen, who died climbing one of the great mountain peaks of the world, and Simon, who died in a helicopter crash in a remote jungle. Both alluded to the risks of what they undertook in their interviews, and exuded passion for their travel. Simon spoke of the importance of exploring the bottom of the ocean, which he argued was still unfinished business and largely unknown. Both were driven to achieve, and talked about not wanting to have regrets later in life. They died doing what they loved best.

Part 2

Imagining Explorers

6   Fiction and the Myth of the Explorer

It is 1862. There is pandemonium at the Royal Geographical Society in London. A bold attempt to cross Africa has just been announced to the assembled members:

> This was a gathering of bold explorers, aged and worn, whom their rest-less temperaments have dragged through the four quarters of the world. Physically or morally, they had practically all escaped from shipwreck, fire, the tomahawk of the Indian, the club of the savage, the torture-stake, and Polynesian stomachs! But nothing could restrain the leaping of their hearts. (Verne, 1862: 165)

Although the setting is real, this is a fictional explorer's account, albeit based on contemporary happenings. Jules Verne's first novel, *Five Weeks in a Balloon*, starts with the excitement of the call to adventure. The hero is Dr Samuel Ferguson, the exemplar of imperial explorers and Victorian character:

> And the doctor entered amid a thunder of applause, and without the least show of emotion. He was a man of about forty, of average height and build ... His expression was cold, his features regular, with a prominent nose, the figure-head nose of a man predestined for discovery. His eyes, very intelligent and gentle, rather than bold, gave great charm to his face. His arms were long, and he placed his feet on the ground in the confident manner of a great walker. The whole person of the doctor exhaled calm gravity. (Verne, 1862: 166)

Ferguson had served in India, then crossed Australia with Stuart, explored the Arctic Circle and trekked in Tibet. After two years of planning he has made public his scheme to fly over Africa. His friend Dick Kennedy tries to dissuade him, arguing:

> Your plan is madness. It's impossible. It's unheard of, beyond all reason ...
> Obstacles, Ferguson answered gravely, are created to be overcome, and as for dangers, who has the confidence to think he can avoid them? There's danger everywhere in life. It may be dangerous to sit at this table or to put your hat on ...
> If you insist on crossing Africa, if you can't be happy unless you do, why not go the ordinary way? ...
> Because up to now every attempt has failed ... Because to struggle against the elements, against hunger, thirst and fever, against savage animals, and still more savage people, is impossible. Because what can't be done in one way ought to be tried in another. (Verne, 1862: 175)

Ferguson's motives are apparently worthy. Africa is a geographical problem that needs to be solved. Whereas others have failed, he hopes to succeed. However, that ambition is questioned when Kennedy continues in his objections:

> He attacked the usefulness of the expedition and its opportuneness ... Africa was certain to be crossed sometime or other, and in a less risky manner ... In a month, six months, before the year was out, some explorer would undoubtedly succeed. (Verne, 1862: 181)

The normally taciturn Ferguson explodes in rage, tellingly revealing what really drives him:

> Is this what you call friendship, you traitor? Do you want someone else to get all the glory? ... don't you realise that my expedition has to compete with others which are already on the way? Don't you know that fresh explorers are making their way towards the centre of Africa? (Verne, 1862: 182)

Imagining the Explorer

Fictional accounts of exploration have long been popular. There was a peak in the late 19th century; mirroring the Golden Age of imperial exploration, a wide range of fictional literature was published dealing with explorers, adventurers and frontier travel. Jules Verne was one who specialised in it; others of note included Henry Rider Haggard, Herman Melville, Joseph Conrad, Rudyard Kipling, Arthur Conan Doyle and Robert Louis Stevenson. Even today, their works are still widely read and regularly adapted for television or film. In addition, modern fiction still dwells on explorers and travellers, not as 19th century imperialists, but as searchers on a quest for things that are lost or hidden. Today's heroes are more likely to be archaeologists, cryptozoologists, scientists, treasure-hunters, or just plain travellers.

In tourism studies, fictional media is generally seen as influencing travel in two ways. The first (and most widely recognised) is that people are drawn to places that they read books or watch films about. Accordingly, for example, a person wants to visit Africa because they have read Jules Verne, Henry Rider Haggard, Ernest Hemingway or Karen Blixen, or seen films based on these works. The second effect is that fiction presents travel as transformative, a process we considered in *Books and Travel* (Laing & Frost, 2012). Here it is not the destination but the journey which influences the person – ideally making them wiser or more mature. Some fictional journeys are constructed as a rite of passage, others as a test of character and courage and others as a quest or meaningful search. Accordingly, it is the inner journey of the self that is more important than the physical journey.

A third factor is also present in fiction. These stories perpetuate the myth of the explorer. They reinforce notions of dress and equipment, of preparation and behaviour. They provide an intertextual template for the would-be explorer traveller. Many of them follow the conventions of the Hero's Journey (Campbell, 1949). There is often a striking and exciting Call to Adventure – whether it be the finding of a map, an ancient relic which provides a clue, or the dying words of some earlier explorer – a plot device that catapults the hero upon the adventure. These are, of course, usually highly imaginative and in strong contrast to the motivations of real explorers. Nonetheless, they capture our attention and reinforce the linkages between exploration, mystery and adventure.

Fictional explorer tales and their authors are intriguingly intertwined with reality. Two examples illustrate this. Henry Rider Haggard wrote *King Solomon's Mines* using the African adventurer Frederick Selous as the template for his hero Allan Quatermain. Selous, in turn, had originally decided to migrate to Africa due to his boyhood fascination with fictional tales of imperial adventures. In addition to Selous, Haggard was friends with explorers Percy Fawcett and Joseph Thomson (Beinart & Hughes, 2007; Grann, 2009). An even more complex literary/explorers' salon developed around playwright J.M. Barrie. In the late 19th century he formed a social cricket team. Its mainstay was the superhuman Arthur Conan Doyle, assisted by Jerome K.

Jerome, A.A. Milne and P.G. Wodehouse (Barrie was unsuccessful in getting his friend H.G. Wells – the son of a notable professional cricketer – to play). The formidable team also included African explorers Joseph Thomson and Paul du Chaillu. Later Barrie became close friends with Robert Falcon Scott and godfather to Scott's son Peter. Near death on his ill-fated polar expedition, Scott penned a letter to Barrie, urging him to look after his family. Barrie took this on as a solemn duty, spearheading a fundraising campaign for the families of all who perished with Scott (Telfer, 2010).

It is notable that it is usually fictional books that are most influential rather than films or television. We cannot just explain this away as generational. Films, after all, have been around for over a century. Rather, it may be explained as being due to the nature of books and the reading experience. Books are an intimate pleasure. Generally we read them alone, quietly, by ourselves. With a book we can escape into a personal world, plunging deeply into our own imagination. Fictional books are closely associated with childhood. They are an antidote to rainy days or periods of illness. In childhood, reading may be ritualised as, for example, in reading or being read to before going to sleep, or in receiving books as presents for birthdays or Christmas. In adulthood, childhood reading and favourite authors become bathed in a nostalgic glow and for many there is great pleasure in revisiting these classics. Reading about explorers and travel has this profound depth, ensuring that it works quite differently from film and television (Laing & Frost, 2012).

For the reader, explorer narratives can be either true stories or fiction. While both have similarities, there are some key differences. While fiction lacks authenticity, it can explore fantasy. The fictional writer has the opportunity to invent unusual group dynamics (such as stowaways, romantic interests rescued en route, or eccentric characters), strange modes of travel and, perhaps most importantly, fantastic places and cultures. The last is apparent when fictional authors allow their adventurers to cross the border from the known to the unknown. In the 19th century, this unexplored territory could be *darkest* Africa, under the sea, even under the ground. In more recent times, we have tended to aim towards outer space. By venturing into the unknown, the author promises the reader an escalating array of outlandish imagined encounters.

In this chapter, we consider a range of fictional explorers. Our choices are purely subjective, either examples that have fascinated us or those we see as representing the fictionalisation of an important element of exploration. We are conscious that our examples are strongly gendered. This highlights an important disconnection. In reality, there were and are female explorer travellers – the 19th century saw, for example, Isabella Bird, Edith Durham and Mary Kingsley. In fiction, they are under-represented, confined really only to such late 20th century creations as Lara Croft and Adèle Blanc-Sec.

The Golden Age of Exploration Fiction

If the period from around 1850 to WWI was the Golden Age of Exploration, it was also the Golden Age of Exploration Fiction. Here we consider three examples in the writings of Jules Verne, Henry Rider Haggard and Arthur Conan Doyle.

The *Extraordinary Journeys* of Jules Verne

Published under the series title of *Extraordinary Journeys*, the works of Jules Verne set the standard for exploration fiction. In addition to marvellous inventions and improbable destinations, Verne's fiction was marked by the theme of the transformative power of travel. In this, it has been argued, Verne was wrestling with issues in his own life, particularly his conservative father's disapproval of his writing and desire that his son settle down to a respectable existence as a lawyer. Accordingly, many of his heroes are outwardly mechanical and unflappable – best exemplified by Phileas Fogg in *Around the World in 80 Days* (1873) – but inwardly are explosive and tormented by their own diminishing humanity. These two contradictory sides of their nature are resolved by travel and exploration. It is their adventures, particularly the unexpected and uncontrollable, which release them from their obsessive side and return them as more settled and happier (Butcher, 1990; Laing & Frost, 2012; Martin, 1990).

Five Weeks in a Balloon (1862) was Verne's first novel and deals directly with the contemporary race to cross Africa. Initially his publisher rejected it as too dry and factual, but then another publisher recommended he revise it with a greater emphasis on the human interactions between the explorers (Martin, 1990). The result was an entertaining mix. As they proceed across Africa, Ferguson lectures Kennedy and his servant Joe on the history of African exploration and the geographical features they observe. Such sections are interspersed with various adventures. They are attacked by monkeys, go hunting and visit a tribal village. Early in their voyage they fill in the gap between southward and northward exploration and discover the source of the Nile. To avoid an anti-climax, the second half of their journey is marked by a series of escalating dangers – of a sort that any exploring expedition might face. They are becalmed in the desert and almost run out of water. Joe gets separated and has to be rescued from hostile tribesmen.

Finally, their balloon starts to leak, providing a suspenseful ending as they limp along the last leg. This structure of a small group experiencing assorted adventures while exploring would be repeated in many of Verne's novels.

In his later works, Verne dwelt heavily on the circumstances that start the imaginary journey. Whereas the voyage in *Five Weeks in a Balloon* was meticulously planned, in other books the Call to Adventure is more fantastic. Verne was particularly fascinated by mysteries and coded messages that unexpectedly throw his heroes out of their everyday routines and into exciting adventures. The nature of these events is such that the heroes are swept along; they are in awe of their new surroundings, but have little control. Such devices clearly struck a chord with his audience, for his most popular works all commence this way.

In *Around the World in 80 Days* (1873), the normally rigid and inflexible Phileas Fogg gets into an argument at his club. A newspaper article claims it is now possible to circumnavigate the globe in 80 days. The argument quickly gets out of hand and Fogg rashly accepts a huge wager to undertake such a trip. In *Twenty Thousand Leagues Under the Sea* (1870), a mysterious something – possibly a sea monster – is wrecking ships. The US government recruits Professor Aronnax to investigate. His boat is attacked and he falls overboard. Rescued, he is now the prisoner of Captain Nemo aboard the futuristic submarine *Nautilus*. In *Journey to the Centre of the Earth* (1864), Professor Liedenbrock finds an old book with a coded message. Deciphering it, he finds it was written hundreds of years earlier by Arne Saknussemm and details a route to the centre of the earth. This stimulates the professor to mount an expedition and follow the message – very much a form of treasure map. In this expedition, Liedenbrock is following Saknussemm's marked pathway rather than exploring on his own account. Interestingly, in film versions this is the one Verne novel where the source of the message is modified for dramatic impact. In the classic 1959 production, the message is carved on a plumb-bob which the dying Saknussemm has thrown into a volcano. He knows that he is finished, but he desperately wants his story to be told and others to follow. In the 2008 remake, the coded message is written within a copy of Verne's novel.

King Solomon's Mines (Henry Rider Haggard, 1885)

On a train to London, Henry Rider Haggard got into an argument with his brother over the merits of Robert Louis Stevenson's *Treasure Island* (1883). The result was a small wager that Haggard could write a better adventure novel. Utilising his experiences as a colonial administrator in South Africa, Haggard quickly wrote the novel that would make his fortune and his name.

Like *Treasure Island*, it starts with a treasure map. The narrator is Allan Quatermain, who has spent most of his life trading and hunting in South Africa. He is approached by Sir Henry Curtis and his companion Captain John Good. Years earlier, Sir Henry had quarrelled with his brother George, who stormed off to the colonies. Sir Henry now seeks a reconciliation and wants Quatermain's help to find his brother. Quatermain has met George Curtis, who was searching for King Solomon's Mines in the interior. This is an old legend; Quatermain had first heard of it 30 years earlier. At one stage he had rescued a dying Portuguese explorer called José Silvestre. In gratitude, he gave Quatermain a map and an account written by his relative in the 16th century. Quatermain has passed these on to George Curtis. Persuaded by a large fee, he agrees to guide Sir Henry in the search for his brother.

After crossing a desert and climbing a mountain range, they arrive in Kukuanaland. A fertile plateau, it is peopled by relatives of the Zulu, who have developed a militaristic society. Also in evidence are colossal ruins, the remnants of an older civilisation. The land is ruled by the cruel Twala and the evil witch Gagool, who distrust the newcomers. When it is revealed that their servant Umbopa is actually the real heir to the throne, the adventurers aid him in overthrowing Twala. As a reward they are taken to King Solomon's Mines. Laden with diamonds, they return through the desert by a different route. At an oasis they find George Curtis. Having broken his leg, he has been stranded.

On one level this is a *boys' own adventure*, with action and derring-do, plot twists and impossible escapes, villains and evil witches. However, on another level the story follows the classical structure of a *katabasis*, literally a descent into hell. Originating with the ancient Greeks, it is common to myths in many cultures and used widely in fiction (examples include *Lord of the Rings* and *Harry Potter*). The classic katabasis involves a long and dangerous journey with six features (Frost & Laing, 2012; Holtsmark, 2001; Laing & Frost, 2012):

(1)   The destination is the realm of the dead or some similarly hellish land.
(2)   The journey is necessary to recover something or find someone.
(3)   A guide is needed. They need to be paid or tricked into helping.
(4)   The realm of the dead is more than dangerous. It is nightmarish and haunted.
(5)   Someone must die as a sacrificial victim.
(6)   The journey changes the hero, either by a striking personal transformation or they receive either a benefit or a curse. The benefit may be seen as equating to the notion of the hero gaining a boon through the journey (Campbell, 1949).

The quest, of course, is to find Sir Henry's brother. Kukuanaland should be a fertile paradise, but it is tainted by a bloodthirsty culture dominated by ritual human sacrifice. King Solomon's subterranean mines are known as the 'Place of Death'. Descending, they are greeted by a ghastly sculpture, 'at the end of a long stone table, holding in his skeleton fingers a great white spear, sat *Death* himself, shaped in the form of a colossal human skeleton, fifteen feet or more in height…the grinning, gleaming skull projected towards us…the jaws a little open, as though it was about to speak' (Haggard, 1885: 266–267). Seated at the table are the bodies of the dead kings of the Kukuanas.

Trickery is needed to survive. Asked by King Twala where they come from, Quatermain replies, 'We come from the stars, ask us not how. We come to see this land.' Twala knows they are mortal, but plays along: 'Ye speak with a loud voice, people of the stars. Remember that the stars are far off, and ye are here' (Haggard, 1885: 144). When sentenced to death, they use their knowledge of a forthcoming eclipse of the sun, to convince the natives they are wizards.

Finally, they force the witch Gagool to take them into the mines, although she plays them false, releasing the trapdoor to seal them in. The sacrifice is made by Foulata, a young native girl, who dies fighting Gagool. The travellers return with three benefits: a ripping yarn, the lost brother and a bag of diamonds. However, Umbopa, now the king, warns them that they must never return. He has travelled widely in exile and does not want white men to come, either as missionaries or traders with rum.

Haggard cleverly interwove into his plot contemporary ideas and stories about Africa that were familiar and appealing to late Victorian readers. These reinforced the notion of the armchair traveller sharing in the exploration of the *Dark Continent*. The quest for George Curtis is akin to Henry Morton Stanley's search for David Livingstone. The mysterious ruins are based on those of Great Zimbabwe, 'rediscovered' by the German explorer Karl Mauch in 1871. At the time there was much discussion about their origins. Many Europeans found it hard to believe that Black Africans could have achieved such constructions, so there was much fanciful speculation regarding their having some Biblical connection. The subterfuge of escaping from capture by seemingly causing an eclipse of the sun may have been inspired by John C. Cremony in *Life Amongst the Apaches* (1869). Cremony and another officer were observing an eclipse of the moon. Some Apaches asked what they were doing with their telescope and Cremony joked, 'we are shooting and killing the moon'. As the moon disappeared, the Apaches panicked and threatened to kill them. Cremony countered by claiming to be a great Medicine Man: 'It is also in our power to restore it to health and strength; but if you harm us or injure our instruments, then the moon must remain dead' (Cremony, 1868: 99–100). Finally, the military organisation and prowess of the Kukuanas linked to the Zulu massacre of a British column at the Battle of Isandlwana in 1879 (this and the Battle of Little Bighorn in 1876 are generally regarded as the only examples of traditional indigenous victories over major Western armies). In turn, intertextuality is apparent in how Haggard's novel has been highly influential on later fiction. *Lord of the Rings*, for example, draws heavily on a journey through the colossal ruins of a forgotten higher civilisation. *Quatermass*, the British Gothic Science Fiction series of the 1950s and 1960s (and precursor to *Doctor Who*) draws the name of its eponymous adventurer/explorer hero from the narrator of Haggard's novel.

The Lost World (Arthur Conan Doyle, 1912)

This is a light-hearted adventure, self-consciously a *book for boys*. The Lost World is a plateau in South America inhabited by dinosaurs and cavemen. Our chief interest here is in the genesis of the expedition. As with the novels of Jules Verne, the hero rapidly moves from the everyday to exciting adventures and the uncertainty and lack of control are part of the appeal for the reader.

The story starts with broad comedy. The hero Malone is a newspaperman. When he proposes to Gladys, he is rejected. She wants a man 'of great deeds and strange experiences ... the glories he had won ... they would be reflected upon me' (Doyle, 1912: 12). She names Burton and Stanley as her heroes; she wants to be like their wives. Rebuffed, he asks his editor for some sort of dangerous mission to prove himself. He is sent to interview the cantankerous and comical Professor Challenger, a chance occurrence that will turn him into an explorer.

In a scene reminiscent of Verne's works, Challenger causes uproar at a meeting of the august Zoological Institute. When a speaker observes that dinosaurs are extinct, Challenger belligerently interrupts, asserting that they still exist in unexplored parts of the Amazon. As his name suggests, Challenger's role is to attack scientific canon. He proposes an expedition to his *Lost World*. Malone seizes the chance and volunteers, as does the experienced adventurer Lord Roxton (a character based on real-life Amazon explorer Percy Fawcett; see Grann, 2009). Completing the quartet is the sceptical Professor Summerlee.

Challenger has discovered the Lost World through a chance encounter with a dying traveller. Again this is reminiscent of an earlier classic – *King Solomon's Mines* – and follows the convention observed by Campbell (1949) of a heroic journey starting by accident. Among the dead traveller's belongings,

Challenger finds a sketch of a Stegosaurus. Reasoning that it has been drawn from life, Challenger deduces the location as a remote plateau. Journeying there, they do encounter a varied mix of prehistoric life.

Returning to London, they attend a packed meeting of the Zoological Institute. Summerlee reports that Challenger was right. Nonetheless, there is opposition. A Dr Illingworth questions the veracity of their account:

> Was this to constitute a final proof where the matters in question were of the most revolutionary and incredible character? There had been recent examples of travellers arriving from the unknown with certain tales which had been too readily accepted ... he admitted that the members of the [exploring] committee were men of character. But human nature was very complex. Even Professors might be misled by the desire for notoriety ... Heavy-game shots [i.e. Roxton] liked to be in a position to cap the tales of their rivals, and journalists [Malone] were not averse from sensational *coups* ... Each member of the committee had his own motive for making the most of his results ... The corroboration of these wondrous tales was really of the most slender description ... I move ... the whole matter shall be regarded as '*non-proven*'. (Doyle, 1912: 197–198)

Although this is a playful book, Doyle has ventured into one of the key controversies of exploration – that of proof. He was writing at a time when a number of claims were being questioned (Conefrey & Jordan, 1998). How did an explorer prove that they had reached the North Pole or the Source of the Nile? What could they bring back?

Challenger has a small number of blurry photographs, but these are ridiculed as easily faked (ironically, within a decade Doyle would ruin his reputation by being taken in by supposed photographs of fairies). The professor offers to show the assembly a photograph of a pterodactyl, leading to the following heated exchange:

> Dr Illingworth: 'No picture could convince us of anything.'
> Professor Challenger: 'You would require to see the thing itself?'
> Dr Illingworth: 'Undoubtedly.'
> Professor Challenger: 'And you would accept that?'
> Dr Illingworth: (*Laughing*): 'Beyond a doubt.' (Doyle, 1912: 200)

Illingworth has fallen for Challenger's trap. He brings in a case which does indeed contain a live pterodactyl: 'the face of the creature was like the wildest gargoyle that the imagination of a mad medieval builder could have conceived ... it was the devil of our childhood in person' (Doyle,

1912: 200). However, Challenger's triumph is short-lived. In the ensuing uproar, the pterodactyl panics and takes to the air. His proof literally flies out the window.

Parodying Explorers

As the explorer archetype became a well-known figure in fiction, so developed the potential for satire and parody. The hero of imperial expansion could be humorously transformed into a self-important and pompous wind-bag. In some cases, part of the fun of imperial ripping yarns was the inclusion of a ridiculous explorer, as in the overblown and hot-tempered Professor Challenger in *The Lost World*. Such comedic fictions pricked holes in the myth, provoking knowing laughter at the explorer's obsession with clothing and equipment and holding up to ridicule the self-obsession and narcissism of the hero. Two examples are worth considering briefly.

Three Men in a Boat (Jerome K. Jerome, 1889)

In this much-loved comic novel, three friends – the narrator, Harris and George – go on a boating holiday up the Thames. The humour comes from escalating exaggeration and mock seriousness. Accordingly, the prepara-tions for the trip are treated as if they were setting out on an exploration expedition to the unknown. Much attention is directed towards having the right outfit:

> The river affords a good opportunity for dress. For once in a way, we men are able to show *our* taste in colours, and I think we came out very natty, if you ask me. I always like a little red ... my hair is a sort of golden brown, rather a pretty shade I've been told and dark red matches it beau-tifully; and then I always think a light-blue necktie goes so well with it ... and a red silk handkerchief around the waist ... Harris always keeps to shades or mixtures of orange or yellow, but I don't think he is at all wise in this ... [for George] the blazer is loud ... the man had told him it was an Oriental design. (Jerome, 1889: 64)

They have so much equipment and provisions that their departure attracts a crowd. 'They ain't a-going to starve, are they', voices one onlooker. Another suggests that with such preparations they could cross the Atlantic. Finally making the specific connection between overpacking and real explorers, a cheeky greengrocer's boy shouts out, 'they're a-going to find Stanley' (Jerome, 1889: 49).

Animal Crackers (play 1928; filmed 1930)

Groucho Marx specialised in playing dubious characters purporting to hold positions of authority. Politicians, doctors and professors were included in his repertoire, so it is not surprising that he would also play an explorer. He is Captain Jeffrey T. Spaulding. Just back from Africa, he is the guest of honour at a party in a swanky country house owned by nouveau riche Mrs Rittenhouse (Margaret Dumont).

Spaulding understands that he is there to play a part, to entertain the rich. Hopefully he will gain funding for other adventures, possibly even his retirement. He makes a grand entrance attired in a pith helmet, boots and jodhpurs – the height of explorer fashion. That night, as a prelude to a musical entertainment, Spaulding is asked to give a lecture on his recent trip to Africa. For Groucho this is an opportunity for a string of one-liners: 'One morning I shot an elephant in my pyjamas. How he got in my pyjamas, I don't know'. However, among the humour there is an insight into how travel to Africa by the rich was conducted, what Cameron (1990) has called *champagne safaris*. Spaulding is not seeking geographical knowledge; rather his lecture is a litany of what animals he tries to shoot. Travel is leisured, 'up at six, breakfasted and back in bed at seven' and 'the champagne must be cold'.

New Finds

From the early 20th century onwards, the direction of explorer fiction changed. Building on the perception that there were few new *places* to discover, authors turned to explorers looking for *things*. Adding drama was the idea that these were hidden. Accordingly, the new heroes were archaeologists or similar.

Murder in Mesopotamia (Agatha Christie, 1936)

Agatha Christie, by her own admission, wasn't terribly absorbed by history when she travelled as a young woman. Her mother took her to Cairo for the winter when she was just 17, and she scarcely ventured forth from the Egyptian capital, eschewing visits to the ruins at Luxor and Karnak in favour of picnics, polo and dances with young men: 'The wonders of antiquity were the last thing I cared to see' (Christie, 1977: 171). This is a surprising lack of curiosity for what would become the exotic backdrop of one of her much-loved detective novels, *Death on the Nile* (1937) and two works set in ancient Egypt – the play *Akhnaton* (1937) and the thriller *Death Comes as the End* (1944).

Christie was always an observant person, with an eye for detail in respect of physical surroundings and people's characters (Morgan, 1984). This was to serve her well as a writer and as the wife of an archaeologist. She delighted in creating stories, as well as travelling to far-away places, as fuel for her imagination but also as a means of escape. In her own words, when travelling:

> You step from one life into another. You are yourself, but a different self. The new self is untrammelled by all the hundreds of spiders' webs and filaments that enclose you in a cocoon of day-to-day domestic life … (Christie, 1977: 307)

Her interest in adventurous travel was piqued in 1922, when she toured the world with her then husband Archie on a mission to publicise the British Empire Exhibition. The couple visited South Africa, Australia, New Zealand and Canada and made time for a detour to Fiji and Hawaii. At that stage, Christie had published a few novels, but was not yet known as the queen of crime fiction. Her marriage to Archie subsequently collapsed, and she made her famous disappearance, being found in a hotel in Harrogate 10 days later, apparently suffering from amnesia. This drama played itself out in the press, after a high-profile police hunt for the missing author (Morgan, 1984). Newly divorced, she found England 'unbearable'. In 1928 she booked a holiday to Jamaica, but cancelled it after meeting a naval officer and his wife over dinner, who talked of being stationed in the Persian Gulf and the delights of Baghdad. Here the author was sent on her journey through a small accidental incident. After that dinner, she made the snap decision to visit Ur, to see the sites where the archaeologist Leonard Woolley was working. Christie was told that the best way to get there was to take the *Orient Express*, which she calls 'my favourite train' (Christie Mallowan, 1946: 23). Her thrill at the thought of rail travel was to inspire her books *The Mystery of the Blue Train* (1928) and *Murder on the Orient Express* (1934).

For Christie, Baghdad held 'all the pleasures of the unknown' (Christie, 1977: 362). She travelled by herself, having decided 'It's now or never. Either I cling to everything that's safe and that I know, or else I develop more initiative, do things on my own' (Christie, 1977: 361). For a woman of her time and social class, it was a bold step. As she noted (Christie, 1977: 379): 'Except for renowned travellers, women seldom went about alone'. However, there were precedents throughout her life: her work during WWI as a VAD and at the dispensary in a hospital, which acquainted her with poisons and the horrors of war wounds; her persistence in hawking her novels around various London publishers until she finally met with interest in her work;

and her risk-taking when she accompanied Archie around the world, and came back penniless.

The decision to visit Baghdad was to transform her life. Christie's time in Ur was deeply satisfying and she realised how much she enjoyed life on a dig – 'the lure of the past' (Christie, 1977: 377). This makes sense for someone enthralled with the *clue puzzle* genre of detective novels (Laing & Frost, 2012), given the similarities between solving a murder mystery and piecing together the stories of antiquity. Christie looked back with regret to her frivolous time in Egypt as a young girl, and made plans to return and to see other parts of Iraq. This time she met her second husband, the archaeologist Max Mallowan, who escorted her around at the insistence of Leonard Woolley. Although 15 years younger than she was, Max liked her no-nonsense manner and ability to deal with crises with equanimity and good humour. She liked his solidity and decisiveness, getting things done, rather than just talking about them. With their shared interests and temperaments, they were the perfect travelling duo. Christie accompanied him on many archaeological digs after they were married, and wrote *Come Tell Me How You Live* (1946), an account of their adventurous journeys through the Middle East. Christie often slept in tents, got used to their vehicle breaking down or getting bogged, wore out her shoes in tramping through archaeological sites and assisted the expedition by taking photographs and helping with the recording and restoration of pottery (Christie Mallowan, 1946; Trümpler, 1999b). She lived up to her second husband's initial assessment of her as a phlegmatic trouper. Christie also used her knowledge of this life as the basis of the plot of *Murder in Mesopotamia*. She wrote it in 1935, at the site of an excavation at Chagar Bazar in Syria.

The story begins with a nurse, Amy Leatheran, engaged to look after Louise Leidner, the wife of the head of an American expedition at the site of an ancient Assyrian city in Iraq. Louise is nervy, beautiful and detested by many of the expeditioners. She is afraid of being killed. We learn early on that Louise is not suspicious of the Arabs – she speaks Arabic and enjoys their sense of humour. In this instance, she reflects Christie's enlightened view of the Arab world, in comparison to many of her contemporaries (Trümpler, 1999c). It is the *Westerners* in this book who take on the role of the *Other* and who arouse Louise's distrust. She has received letters that warn of impending tragedy: 'Death is coming very soon'. Were they from her first husband, who had threatened her in the past but was now believed to be dead? Or was the source closer to home?

Nurse Leatheran arrives at the expedition house to find a 'strained atmosphere'. She is not a keen traveller, and complains about the roads, the dirt, the chaos and the delays she encounters: 'not romantic at all like you'd think from the *Arabian Nights!*' (Christie, 1936: 13). Her dislike of her surroundings is a useful plot device, in that she is not interested in the activities of the expedition ('Messing about with people and places that are buried and done with doesn't make sense to me' – Christie, 1936: 67). Accordingly, she is a disinterested bystander or onlooker, able to size people up, and helps the detective Hercule Poirot solve the case.

The expedition headquarters functions as a *locked room* (Cholidis, 1999), a common denominator in most Christie plots (Hardyment, 2000). Built around a courtyard, the building has barred windows looking out onto the surrounding countryside and its only entrance is locked at night and guarded by day. The inevitable happens, and Louise Leidner is found clubbed to death in her bedroom one afternoon. It appears to be an inside job, and each member of the expeditionary team is suspected of having a hand in the deed.

Poirot arranges for the suspects to be gathered together – his theatrical signature move. He tells them that they will begin on 'A journey into the past. A journey into the strange places of the human soul' (Christie, 1936: 301) and then quotes the Arabic phrase used before commencing travel: 'In the name of Allah, the Merciful, the Compassionate'. Even prosaic Nurse Leatheran is moved by his words, and conjures up in her mind a vision of 'merchants with long beards – and kneeling camels – and staggering porters carrying great bales on their backs held by a rope round the forehead – and women with henna-stained hair and tattooed faces kneeling by the Tigris and washing clothes' (Christie, 1936: 301). This is the first inkling that her time in the Middle East has transformed her and given her an appreciation of the local culture – 'like a piece of fusty old stuff you take into the light and suddenly see the rich colours of an old embroidery' (Christie, 1936: 302).

The twist in the tale is that Louise's current husband, Eric, is actually her first husband, who was supposed to have been shot as a spy after his wife reported his traitorous actions to the US authorities during WWI. Infatuated with his wife, he marries her years later and maintains his disguise as an archaeologist. Eric kills Louise by dropping a heavy millstone on her from above, after she opens her bedroom window. His motive for killing his wife is not retribution for her past betrayal, but jealousy over her love affair with Richard Carey, a young architect. The murder weapon is an archaeological object, a deft use of the type of artefact that Christie would have been familiar with from her time at various digs (Cholidis, 1999).

As a thriller, the book is a little tired and the plot overly fantastic. But its appeal lies in its unusual setting, the bringing to life of an archaeological expedition, both the mundane activities and the thrill of discovery, and the sympathetic treatment of the East. It is the Westerners who are depicted in some cases as duplicitous and morally suspect, as well as narrow minded, despite their travels. Father Lavigny is revealed to be a French jewel thief and Nurse Leatheran is sanctimonious, snobbish and racist, with her comment that Mrs Mercado 'might have had what my mother used to call "a touch of the tar-brush"' (Christie, 1936: 48). Louise Leidner, outwardly elegant and enigmatic, is also cruel and self-centred, in a thinly disguised portrait of the magnetic and mercurial Katherine Woolley, wife of the expedition director at Ur (Cholidis, 1999; Mallowan, 1977). *Murder in Mesopotamia* depicts the allure of the Other, yet suggests that human nature is often the same the world over, no matter how exotic the surroundings.

The Hunter (2011)

About a decade ago, Warwick went to a lecture by Robert Paddle at the Museum of Victoria. Paddle had researched the history of the extinction of the Tasmanian Tiger or Thylacine, a wolf-like marsupial. The last Thylacine in captivity had died in 1936, although from time to time there are reported sightings. Come question time, the good-sized audience asked Paddle if he had seen a Thylacine and whether he believed there were still remnant populations in the Tasmanian wilderness. His sceptical responses did nothing to diminish the enthusiasm of the crowd. Finally, somebody asked him what he thought the Tasmanian National Parks Service would do if they discovered a living specimen. He replied that he thought they would probably keep it quiet and continue to deny its existence to discourage hunters and tourists. Aha, gasped the questioner, that is exactly what they are doing!

Such an idea underlies the film *The Hunter*. Martin (Willem Dafoe) is hired by military biotech company Redleaf to catch a Thylacine. They want it for its toxic venom, which they see as highly valuable. In reality, the Thylacine has no such venom, although if anyone found one, it would be the scientific discovery of the century and would indeed be worth millions of dollars. In this sense, the premise of the film – that a Thylacine is inconceivably valuable – is correct. Periodically, there are expeditions searching for it, funded by individual enthusiasts, universities, government agencies, zoos and media companies (Paddle, 2000). Perhaps the most famous was by *National Geographic*. With no footage of the Thylacine, their documentary was forced to focus on the Tasmanian Devil, which was picked up by Warner Brothers as Bugs Bunny's adversary, Taz.

Martin's search parallels that of many explorers. He works alone, a lone wolf searching for a lone wolf. Jack (Sam Neill) offers to be his guide and is persistent in wanting to be involved, but Martin rejects his help. Martin thinks that his previous experience puts him at the elite level, but quickly finds that he is struggling in the extreme environment of Tasmania's high country. His quest will take longer than he expects.

Like many explorers, Martin relies on those who have gone before. Trawling the internet, he finds film of a Thylacine in a zoo, but none of them in the wild. He follows in the footsteps of Jarrah, an enigmatic figure who may have discovered the Thylacine's lair, but has now disappeared. Martin works through notes left by Jarrah, searching for clues. Finally, it is Jarrah's son who provides the vital information. Jarrah is the real discover, but as he died before he could reveal his knowledge, he will never get any credit. Martin is then a *rediscoverer*; his exploration is in following Jarrah's trail and finding what he found.

Nor is Martin alone in the wilderness. Jack tells him that they get plenty of tourists in four-wheel drives, dreaming that they could get lucky and discover the fabled animal. Martin sees signs of others and occasionally hears gunshots. Are these competitors? He starts to get paranoid, wondering if someone else will beat him to his goal. One day he finds a camera connected to a motion sensor. He takes the memory card, deliberately sabotaging his unseen rival. As his search lengthens, he finds that Redleaf has become impatient with him, hiring another ruthless explorer to take on the task. The disillusioned Martin muses to himself that the search for the Thylacine is endless; there will always be individuals looking for it, whether it exists or not.

The Explorer's Legacy

For all the adventure, heroics and humour in the fictional explorer narratives, there is also a strong dark undercurrent. Exploration often leads to conquest, exploitation and misery, with a more developed society discovering and taking control of a weaker one. We conclude this chapter with the darkest of the explorer fictions.

Heart of Darkness (Joseph Conrad, 1899)

Marlow has always been an explorer, recounting to his friends:

> When I was a little chap I had a passion for maps. I would look for hours at South America, or Africa, or Australia, and lose myself in all the glories of exploration … there were many blank spaces … I would put my finger on it and say, When I grow up I will go there. (Conrad, 1899: 9)

He becomes a sailor – a steamboat captain – and travels through the Far East and the Pacific and Indian Oceans. However, it is Africa that calls out to him. The pull is too great, so he takes a job with a European trading company and heads for the Belgian Congo. What he finds almost immediately disappoints him. The natives have been enslaved. Chained and maltreated, death is everywhere. His fellow employees are indifferent. He sneeringly refers to them as *pilgrims*, journeying to Africa greedy for instant wealth, but with no humanity or desire to work for advancement. He travels with a group fantastically called the Eldorado Exploring Expedition:

> Their talk ... was the talk of sordid buccaneers: it was reckless without hardihood, greedy without audacity, and cruel without courage; there was not an atom of foresight or of serious intention in the whole batch of them, and they did not seem aware that these things are wanted for the work of the world. To tear treasure out of the bowels of the land was their desire, with no more moral purpose at the back of it than there is in burglars breaking into a safe. (Conrad, 1899: 42)

Marlow and the pilgrims go up a river to find Kurz. Regarded as the very best of the company agents, there is an expectation that Kurz will unlock the riches of the interior. Instead, they find he has *gone native*, carving out a petty principality for himself. His African princess is a 'wild and gorgeous apparition of a woman'. Adorned with numerous items of jewellery, she 'must have had the value of several elephant tusks upon her'. Marlow finds her 'savage and superb, wild-eyed and magnificent' (Conrad, 1899: 85–86). However, Kurz has been upriver for too long. He is dying, broken by fever and killing: 'his soul was mad. Being alone in the wilderness, it had looked within itself and by heavens! I tell you it had gone mad' (Conrad, 1899: 94). As he dies, Kurz's last words are 'The horror! The horror!' (Conrad, 1899: 98).

The disillusioned Marlow returns to Europe. He has Kurz's papers and company agents try to get them, still believing that Kurz had discovered great wealth. Marlow visits Kurz's fiancée. She idolises Kurz, portraying him as a great explorer and empire builder cut down in his prime. Guiltily, Marlow agrees. He cannot tell her the truth. When she asks what his last words were, Marlow readily lies that it was her name.

Conrad's novel was intended as a damning indictment of Europe's opening up of Africa. Discovery does not lead to churches, schools or hospitals, but rather to violence and destruction. The Europeans introduce a brutal regime purely to extract wealth. They have no higher purpose. In Europe there is the pretence of civilising colonisation; the truth is that this is a lie. Kurz, who everyone expects to do great things, lives that lie. The word Kurz can be translated from German to mean *cut short* or *lessen*. For Conrad, that is the legacy of exploration in Africa.

7   Desert Island Castaways

Being stranded on a desert island is a well-used trope within adventure fiction. Faced with such a predicament, fictional explorers resort to the Message in a Bottle. Second only to the Treasure Map, it provides a recurring example of Campbell's (1949) *Call to Adventure* for the hero. As discussed in Chapter 2, real-life explorers rarely stumble into their adventures, but in fiction and mythology the accidental beginning of a journey is more common and the finding of a message in a bottle is a random occurrence to change the hero's pathway. The found message functions in two ways. The first – as in real life instances – is that it is a cry for help. The writer is lost. Stranded in some remote place, they are barely surviving, but cannot return. They need help. The use of the bottle is a desperate, one in a million, attempt at calling for a rescue party. The second function is more a literary device. The explorer is stranded and knows there is really no hope of rescue. Instead of calling for help, they utilise the message in the bottle to tell their story, either describing the new lands they have experienced and explored, or warning of its dangers.

Two examples illustrate how the message in the bottle is used to convey the stranded explorer's knowledge back to home. The first is by Jules Verne in his *In Search of the Castaways* or *The Children of Captain Grant* (1867–1868 and filmed in 1962 and 1996). Verne loved mysterious coded messages that his heroes had to break in order to get the story moving, as in *Journey to the Centre of the Earth* (discussed in Chapter 6). *In Search of the Castaways* begins when Lord and Lady Glenarvan find a champagne bottle inside a shark. It contains three messages, in English, French and German, respectively. All are incomplete, having been spoiled by seawater seeping in. Deducing the text is the same, just in different languages; they piece together enough parts of the message to mostly understand it. It is a call for help from the shipwrecked Captain Grant. They advertise in the newspaper for relatives and after meeting his destitute children decide to stage a search around the globe for the stranded Grant. However, they face a problem. The location of the wreck is incomplete. Latitude (37 degrees) is included, but the longitude has been rubbed out. This requires their expedition to traverse the globe along 37 degrees south, taking in the barely explored regions of Patagonia, Tristan da Cunha, Australia and New Zealand.

The second example transfers this concept to outer space. Pierre Boulle's *Planet of the Apes* (1963) starts with two adventure travellers – Jinn and Phyllis. In a small rocket, they wander through the solar system like two modern-day backpackers. Occasionally they come across interesting objects floating in space. One day they find a bottle with a message. It is from the explorer Ulysse Mérou. It starts with the warning: 'I am confiding this manuscript to space, not with the intention of saving myself, but to help, perhaps, in averting the appalling scourge which is menacing the human race' (Boulle, 1963: 10). He recounts his adventures on the eponymous planet. At the end Phyllis is inclined to believe the story, but Jinn convinces her that it cannot be true. They continue their journey.

The recurring motif of the stranded adventurer is a key subcategory of explorer fiction. It is a striking example of intertextuality, as both explorers and novelists cite castaway fiction as influential. In Verne's *Five Weeks in a Balloon*, the youthful Samuel Ferguson has 'his imagination kindled by the reading of bold enterprises and exploration by sea'. One of his favourites is *Robinson Crusoe*:

> What absorbing hours he spent with him on the island of Juan Fernandez! The ideas of the solitary sailor frequently met with his approval, but at times he disputed his plans and schemes. He himself would have acted differently, perhaps better, certainly as well. But one thing is sure; He would never have left that joyous island. (Verne, 1862: 167)

What lies beneath the appeal of the desert island? Is it simply the suspense of whether or not the hero will find a way back home? Is the interest in the practical side of how they survive? Is it the challenge of being tested by adversity and unknown dangers? Is there a romantic dream – as illustrated by Verne's Ferguson – that it promises such an idyllic escape from the everyday that one would not want to leave? In this chapter, we aim to examine these questions through a discussion of seven fictional castaway stories. These are tales of lost explorers and adventurers, of travel that has gone wrong and where the narrative now focuses on the difficulty of the *Return*. We start with *Robinson Crusoe*. The other six are essentially retellings of this classic tale, though all with new variations. Three of the stories were written during the Golden Age of exploration and three are modern, with one of these set in outer space.

Robinson Crusoe (Daniel Defoe, 1719)

Robinson Kreutznaer is a wanderer. Although living in England, his father is a German merchant and their surname gets shortened to Crusoe. He does not really fit in, 'being the third son of the family and not bred to any trade, my head began to be fill'd very early with rambling thoughts' (Defoe, 1719: 5). In defiance of his father, he runs away to sea. His resulting adventures form a moral tale, warning of what happens to those who disregard their fathers' wisdom. His first voyage is abruptly terminated by his boat sinking in a storm. The captain warns him, 'young man, you ought never to go to sea any more, you ought to take this for a plain and visible token that you are not to be a seafaring man' (Defoe, 1719: 13). Captured by pirates, he escapes and makes his way to Brazil, where he develops a successful plantation. However, once again the urge for adventure leads him to go to sea. As he reflects, 'I was still to be the wilful agent of all my own miseries ... procured by my apparent obstinate adhering to my foolish inclination of wandering abroad' (Defoe, 1719: 30).

Once again a storm wrecks his ship and he is the only survivor. He now spends 28 years as a castaway on a small island off the South American coast (Defoe's story is generally believed to be based on the adventures of Alexander Selkirk, who was marooned on a Pacific island for four years). At first, Crusoe is distraught and depressed. Reflecting on his predicament, he draws up a balance sheet: 'I stated it very impartially, like debtor and creditor, the comforts I enjoy'd against the miseries I suffer'd'. Accordingly, he writes on the evil side of his ledger, 'I am cast upon a horrible desolate island, void of all hope of recovery' and then balances this on the good side with, 'But I am alive, and not drown'd as all my ship's company was' (Defoe, 1719: 50).

Crusoe's adventures are divided into three parts. In the first, he salvages what he can from the wreck and sets up house on the island. This practical aspect of survival is one of the common tropes of most castaway stories, many of which consciously reference this novel. To recount a plausible yet exciting story, there must be an accounting of what the lost adventurer has – food, weapons, books, tools – and which essentials are absent.

The second part of the book is an unnerving and paranoid account of Crusoe's growing fear that he will be discovered and killed by cannibals. This part covers the bulk of his time on the island, yet it is covered in the smallest number of pages. The effect is disconcerting. Crusoe seems consumed, almost mad, with fear. Nothing else seems important to him. In contrast, the third part sees Crusoe transformed into an action hero. First, he rescues Friday from the cannibals. Then he rescues Friday's father and a Spanish castaway. Finally, an English boat arrives, but the wary Crusoe hangs back. Something is wrong. When he finds that the crew has mutinied, he leads a successful counter attack. His reward is to be returned to England. This third part is completely different from the rest of the book, but serves the purpose of bringing the adventure to a close. Crusoe is never going to get off the island by himself; he needs an external party to effect his rescue.

The Swiss Family Robinson (Johann David Wyss, 1812)

The family's name is not Robinson. The father, a Swiss pastor, narrates the story, but never tells us his name. Apart from him, there are his wife and four young sons. Castaways, they are a Swiss family living *like* Robinson Crusoe – and they just happen to have that book with them. Wyss, the author, was a Swiss pastor, with a family of boys. There is a vicarious element to the story. Wyss is telling how he would behave if stranded, if there were a *Wyss Family Robinson*.

While emigrating to Australia, their ship is wrecked. Their location is never named, although it seems to be either northwestern Australia or the Indonesian Archipelago. Wyss never visited this part of the world, so there is a degree of invention in what he describes. Like Robinson Crusoe, the castaways are able to salvage a great deal from the vessel, which was stocked with a wide range of equipment for selling to the colonists. Accordingly, they benefit from having livestock, seeds, fruit trees, weapons, clothing and tools. This is then supplemented with the natural resources they find. Hardly worrying about rescue, they continue with their plan of setting up a farming colony in a new land.

The story is episodic. The work of salvage, building houses, planting crops and tending to animals is alternated with discoveries. Wherever they are, it is a curious place. There are monkeys, jackals and flamingos (which they make pets of) and kangaroos (which they eat) and a platypus. The climate is temperate enough for them to grow cherries and walnuts, but they also enjoy the produce of sago palms and coconuts.

The father is a patriarchal leader. He makes all the decisions and takes every opportunity to lecture his family. Through the father, Wyss is advocating the educational philosophy of his fellow Swiss, Jean-Jacques Rousseau (1712–1778). Rousseau's writings emphasised the desirability of a family unit being connected to nature and of a practical education linked to technical crafts. This view dominates the book. The father has little interest in being rescued; the resource-rich island provides him with the materials and closed environment for his social experiment. He constantly refers to knowledge he has gained through reading or visiting museums but, as he stresses, what is most valuable is practical knowledge. He educates his family through a combination of his encyclopaedic memory and experiential learning. A distinctive aspect of this approach is that they construct a *cabinet of curiosities*, a miniature museum holding the stranger items they come across. These include a stuffed platypus (seemingly the only creature the father has never read about), a turtle and a boa constrictor. Such a construction would have appealed to the 19th century audience, this being a period distinguished by collecting and classifying.

Much of the book is taken up with detailed accounts of building or of discovering and processing raw materials. The link between domesticity and the discovery of useful items is illustrated time and time again. For example, in this passage, the father has been stripping bark off trees to roof his house:

We made another agreeable discovery: my wife took up the remaining chips of the bark for lighting a fire … we were surprised by the delicious aromatic odour, which perfumed the air. On examining the half-consumed substance, we found some of the pieces to contain turpentine, and others gum-mastic … the instinct of our goats, or the acuteness of their smell, discovered for us another pleasing acquisition: we observed with surprise, that they ran from a distance to roll themselves on some chips of a particular bark which lay on the ground and which they began to chew and eat greedily … the chips were cinnamon … This new commodity was certainly of no great importance to us, but we regarded it with pleasure, as it might assist to distinguish some day of rejoicing. (Wyss, 1812: 274–275)

The father follows the conventions of the scientific explorer, providing a template for young readers as to how they should proceed in the future. He has a stock of useful knowledge and applies this by further experimentation and observation. Most importantly, he records his newly gained knowledge and collects specimens of the unusual. There is none of the paranoid fear of *Robinson Crusoe*.

As the episodes of discovery continue, there is a tendency for them to become more outlandish. Towards the end of the book, they discover coffee, chocolate and tea in quick succession. Naturally, the father knows how to process all of them. They also encounter lions, wolves, elephants, hippopotami and orangutans. Another castaway – Emily – is found and joins them. After 10 years a British ship finds them. Some colonists aboard are so impressed with their achievements that they decide to settle rather than continue on to Sydney. The father and mother decide this is now home and they opt to stay. So do two of the sons, whereas the other two leave – not to return to Europe – but to see the world.

The Coral Island (R.M. Ballantyne, 1858)

Nearly 50 years on, this starts as a lighter-hearted version of the Swiss Family Robinson. Deliberately changing many of the premises, it holds up much better for the modern reader. First and foremost, its narrator is a teenage boy – Ralph. In the first line he introduces himself and a very different mood: 'roving has always been, and still is, my ruling passion, the joy of my heart, the very sunshine of my existence' (Ballantyne, 1858: 9). In focusing on youthful characters, there is a greater sense of engagement, an approach that would become the staple of children's adventure fiction, as seen in *The Adventures of Tom Sawyer* (Mark Twain, 1876) and *Treasure Island* (Robert Louis Stevenson, 1883).

Ralph, Peterkin and Jack are teenage shipmates. Wrecked on a coral island in the Pacific, they are on their own. There is no father figure to instruct them or impart moral messages. On their first day, the older Jack channels the Swiss pastor, explaining, 'I have been a great reader of books of travel and adventure all my life, and that has put me up to a good many things that you are, perhaps, not acquainted with'. The response from Peterkin is immediate:

> Oh, Jack, that's all humbug. If you begin to lay everything to the credit of books, I'll quite lose my opinion of you. I've seen a lot o' fellows that were *always* poring over books, and when they came to try and *do* anything, they were no better than baboons! (Ballantyne, 1858: 39)

Jack regains his dignity by demonstrating that green coconuts have the sweetest milk and this was something he had read in a book. Nonetheless, it is a clear message that this will be a very different castaway story.

A further difference comes in what they can salvage from the wreck. The Swiss family completely stripped their wreck over weeks. In contrast, these boys only have what is in their pockets. The inventory is dire: a small axe, a broken pen-knife, a telescope and some cord. There is a knowing wink to the reader as the disconsolate Ralph exclaims, 'if the ship had only stuck on the rocks we might have done pretty well, for we could have obtained provisions from her, and tools to enable us to build a shelter, but now – alas! alas! we are lost!' A smiling Jack responds, 'Lost! Ralph? Saved, you should have said'. Peterkin then chimes in that it is 'the best thing that ever happened to us, and the most splendid prospect that ever lay before three jolly young tars' (Ballantyne, 1858: 27). The tone is set; this will be an adventure.

Life on the island is idyllic. The boys swim, dive and fish. There are coconuts, wild pigs and plenty of seafood. Their limited tools and experience are sufficient for all their wants. Contemplating building a house, they decide they do not need one. A purposeful exploration of the island reveals the skeleton of a lone castaway who preceded them. This may be their fate, but they live for the moment. Life is a holiday; rescue can wait.

In the second half of the book, the tone changes. Pirates come ashore and kidnap Ralph, a neat reversal of *Robinson Crusoe*. As one of their crew, he sails around the Pacific, visiting other islands. The South Pacific is in transition as Europeans arrive – either as missionaries, traders or pirates. While not specifically stated, the picture is of traditional societies under great strain, resulting in bloody feuds and raiding. Ralph witnesses a volcanic eruption and makes friends with the fearsome Bloody Bill, the only pirate with a semblance of humanity. They arrive in Feejee (Fiji), aiming to harvest a cargo of sandalwood. As they negotiate with the Fijians, Ralph wanders around, recording their culture. He is fascinated with 'surf swimming':

> For some time the swimmers continued to strike out to sea, breasting over the swell like hundreds of black seals. Then they all turned, and, watching an approaching billow, mounted its white crest, and, each laying his breast on the short flat board, came rolling towards the shore, careering on the summit of the mighty wave, while they and the onlookers shouted and yelled with excitement . . . a few, who seemed to be the most reckless, continued their career until they were launched upon the beach, and enveloped in the churning foam and spray. (Ballantyne, 1858: 307)

Beneath this carefree image, there are violent tensions working away. Annoyed with delays, the pirate captain plans a night attack to teach the natives a lesson. Ralph and Bill make a decision to warn the islanders. All the pirates are killed. Bill, mortally wounded, expresses regret for his wicked life. Now reunited, the boys assist the missionaries. Captured and heading for human sacrifice, they are saved in the nick of time when the Fijian chief is converted to Christianity.

For the modern reader, the second half is too melodramatic and overburdened with Victorian moralising. The first half, providing an alternative version of the *Swiss Family Robinson*, is far more engaging. A *Ripping Yarn* written specifically for children, we can see how it influences later works such as *Treasure Island* and *Peter Pan*.

The Admirable Crichton (J.M. Barrie, 1902)

J.M. Barrie was walking with his friend Arthur Conan Doyle, when they started kicking around the idea of what would happen if a man and his servant were marooned on a desert island (Telfer, 2010). Barrie took it up and developed it into a play. The result was a clever comedy about role reversal.

The First Act introduces Lord Loam and his butler Crichton. Loam has modern ideas about equality, forcing his family to serve morning tea to his servants once a month. Crichton is deeply conservative and disapproves. Loam lectures him, 'Can't you see, Crichton, that our divisions are artificial, that if we were to return to Nature, which is the aspiration of my life, all would be equal'. Crichton, with permission, responds that 'there must always be a master and servants in all civilised communities ... for it is natural, and whatever is natural is right' (Barrie, 1902: 27–28).

Lord Loam starts out for a few months' holiday on his yacht. Along with him are: his daughters Mary, Agatha and Catherine; guests (and potential suitors) Ernest Woolley and the Reverend John Treherne; and Crichton and the maid Tweenie. The ship is wrecked and the passengers are stranded on an island. Faced with this challenge, the aristocrats are hopelessly impractical. While Crichton constructs a shelter, Ernest engages in the important task of writing a message and putting it in a bottle. Lord Loam has read about what to do in such a predicament, although nothing he tries works:

> I remembered from the *Swiss Family Robinson* that if you turn a turtle over he is helpless. My dears, I crawled towards him, I flung myself upon him ... the senseless thing wouldn't wait ... And then those beastly monkeys. I always understood that if you flung stones at them they would retaliate by flinging coco-nuts at you. Would you believe it, I flung a hundred stones, and not one monkey had sufficient intelligence to grasp my meaning ... I tried for hours to make a fire. The authors say that when wrecked on an island you can obtain a light by rubbing two pieces of stick together. The liars! (Barrie, 1902: 65)

The Third Act takes place two years after the shipwreck. Crichton is now the master, known as the 'Guv' – a direct reference to *Robinson Crusoe*, where the hero is given that title after he defeats the mutiny. The others willingly take their orders from Crichton, the natural leader. Like the Swiss Family Robinson, they've built houses and planted crops. Lady Mary has become a skilled huntress. Dressed in skins, she reminds us of Barrie's future creation of Peter Pan, who in theatrical productions is always played by a girl. Crichton and Mary announce they are to marry. All are happy. Then a rescue boat appears.

The Final Act plays out in England. They have all returned to their former roles. Lady Mary, still somewhat tomboyish, is going to marry Lord Brocklehurst. Their history on the island has been rewritten, with Crichton cast as a minor player. Ernest's book of their adventures is a bestseller, one reviewer remarking, 'from the first to last of Mr Woolley's engrossing pages it is evident that he was the ideal man to be wrecked with, and a true hero' (Barrie, 1902: 120). They reflect that their time on the island was like a dream. Crichton gives in his notice and leaves. There is no joyous return home for this heroic traveller.

The Admirable Crichton departs from the previous castaway tropes in two significant ways. First, the survivors represent a more diverse social group than in earlier accounts. They are not just a family or a trio of youths. There is a family, but also friends and servants. They arrive with strong existing relationships. Lord Loam is the natural patriarch. Ernest and John are subservient, but equal socially. And the servants are conditioned to obeying orders. These existing social structures break down almost immediately. As Crichton has foretold, it is natural for there to be leaders and followers. Faced with a fight to survive, roles will have to be different. At the heart of Barrie's reversal is the notion that, while gentlemen lead, it is servants who do the hard practical work. This is a division that sits uneasily with many explorers and adventurers. How much of what they achieve is due to their servants and guides? Reinforcing all of this, when they return to England it is important that a book be published telling their story. However, in this case, the account is written by one of the gentlemen, who rewrites their history to make himself the hero.

The second departure concerns sex. The earlier adventure stories are dominated by males, with practically no females. In stark contrast, the eight castaways in Barrie's play are evenly divided into four males and four females. After two years on the island they are ready to form couples. While specifically referred to in the play, in the 1957 film version this becomes the dominant theme. In an idyllic tropical paradise, romance becomes a priority. This theme is also dominant in the novel *The Blue Lagoon* (Henry De Vere Stacpoole, 1908). In this, a young girl and boy are marooned, so naturally they fall in love.

On the island of *The Admirable Crichton*, the two most desirable mates are the servants. Lady Mary, the most dynamic of the sisters, falls for Crichton. The three male gentlemen are all interested in Tweenie. The social conventions of Britain are abandoned, although the playful Barrie has them rapidly back in place once the adventurers return home. Over a century after it was written, such an observation regarding the rapid return to social roles seems easily applicable to many modern-day travellers.

Darker Modern Narratives

Although the seven seas are now fully explored, the castaway narrative is still with us – demonstrating the difficulties and perils of the hero's return. This more recent fiction puts new twists on tales of survival and adventure. As seen in the next three stories, there is a stronger emphasis on crisis. Darker tales, these castaways face death, their leaders are found wanting and it is their companions who pose the greater danger.

Lord of the Flies (William Golding, 1954)

As the world descends into nuclear war, a group of English schoolboys are marooned on a tropical island. This is Ballantyne's *Coral Island* revisited (Nadal, 1994). Even the two main characters – Jack and Ralph – share the same names as the heroes of the 19th century classic. For the boys, the island is an opportunity for the sort of adventures they have only previously read about. Ralph finds a conch shell and blows it to call all the boys to him. Taking charge, he addresses the assembled boys, assuring them that they will soon be rescued:

> 'While we're waiting we can have a good time on this island.'
> He gesticulated wildly.
> 'It's like in a book.'
> At once there was a clamour.
> 'Treasure Island –'
> 'Swallows and Amazons –'
> 'Coral Island –'
> Ralph waved the conch.
> 'This is our island. It's a good island. Until the grown-ups come to fetch us we'll have fun.' (Golding, 1954: 38)

Their first task, as they know from *The Coral Island*, is to circumnavigate the island. Immediately they settle into the tropes of the explorer narrative:

> 'This is real exploring,' said Jack. 'I bet nobody's been here before.'
> 'We ought to draw a map,' said Ralph, 'only we haven't any paper.' (Golding, 1954: 29)

The odd one out is Piggy (his name is similar to Peterkin, the third boy in *The Coral Island*, and they both share an aversion to swimming in the lagoon). He wants normality and takes on the role of a teacher gathering names at assembly. He voices the idea that there might be no rescue party:

> 'Didn't you hear what the pilot said? About the atom bomb? They're all dead.' ...
> 'They're all dead,' said Piggy, 'an' this is an island. Nobody don't know we're here.' (Golding, 1954: 14–15)

Their temporary society quickly fractures. Both Ralph and Jack want to be the leader. Ralph is determined on keeping a signal fire lit, while Jack is focused on wild pig hunting. Two alpha males, they are so obsessive they cannot compromise, nor co-operate. They compete for recruits, Jack proclaiming 'who'll join my tribe and have fun?' (Golding, 1954: 166). The fire goes out and Piggy laments:

> Just an ordinary fire. You think we could do that, wouldn't you? Just a smoke signal so we can be rescued. Are we savages or what? Only now there's no signal going up. Ships may be passing. Do you remember how we went hunting and the fire went out and a ship passed by? (Golding, 1954: 188)

The crisis of command degenerates into violence. Piggy and another boy are killed. Jack and his hunters try to kill Ralph. Hunting him, they set fire to the island to smoke him out. Their shelters, fuel supply and fruit trees are destroyed. All hope is gone for Ralph, when a naval officer appears. He has seen the smoke and come to the rescue.

Looking at the boys in war paint with spears, the officer enquires, 'Fun and games ... What have you been doing? Having a war or something?' Warming to his joke he asks, 'Nobody killed, I hope? Any dead bodies?' He is startled when Ralph responds that two are dead. Quickly, the officer realises that this is not a joke; something serious has happened. He is appalled that they do not even know how many boys were on the island. He admonishes Ralph: 'I should have thought that a pack of British boys – you're all British aren't you? – would have been able to put up a better show.' Just to emphasise that he too knows his explorer narratives, he continues that he means a 'Jolly good show. Like the Coral Island' (Golding, 1954: 221–223).

Widely taught in schools, this novel was beloved by teachers as it showed that students would quickly degenerate into savages without proper discipline and guidance. However, even easily distracted schoolchildren picked up the true underlying messages. How could adults blame the children for their fighting, when the world had so readily slipped into a destructive atomic war? This is reinforced by it being the navy that rescues them. Furthermore, the boys are the cream of the English school system (it is disturbing that the authorities appear to have only sent the upper class to safety). The imposed discipline and rituals of their schooling were of no help on the island. Rather than leaders, it has created bullies, a new generation primed for war.

Lost in Space (television 1965–1968)

The exploration of space opened a new frontier for reworking stories of marooned castaways. The television show *The Twilight Zone* featured an episode *I Shot an Arrow* (1960) in which a rocket crashes on an inhospitable planet. Arid, fiercely hot, like nothing on earth, the astronauts quickly fall into utter despair. This is not the tropical island of *Robinson Crusoe* and other adventures. Ultimately, only one survives. Near death, he stumbles upon an artefact. It is a signpost to Las Vegas – they have crashed in Death Valley. The screenwriter was Rod Serling and he would use a similar ending for another castaway film in *Planet of the Apes* (1968). Death Valley was the location for the film *Robinson Crusoe on Mars* (1964) (Figure 7.1). This was a straightforward shift of the classic story from the Caribbean to the Red Planet. A scientific expedition crashes; there is one survivor. He must find a way to survive and battle isolation. He even rescues an alien who he calls Friday. It

Figure 7.1 The unworldly landscape of Death Valley
Source: W. Frost.

was not as successful as it deserves to be, perhaps because it stuck too closely to a very well-known story.

Lost in Space also updated a classic, placing it in the future. Producer Irwin Allen initially wanted to call this series *Space Family Robinson*, but was thwarted when it turned out there had previously been a cartoon series by that name. The series is fondly remembered for the interplay between Dr Smith (Jonathan Harris), Will (Billy Mumy) and the Robot. At times it was cartoonish, with Smith camp and cowardly and an increasingly bizarre range of aliens encountered on their voyage. However, the beginning was far more dramatic and dark and it is the first six episodes that we focus on here. These form a complete story arc, introducing the characters and their predicament, and are distinguished by much higher production values than later episodes.

It is 1997 and the world is on the edge of an ecological catastrophe. Faced with massive overpopulation and rapidly declining resources, it is a dark Cold War vision of a not at all bright future. The only hope is a bold plan by the USA to send surplus population to an earth-like planet orbiting Alpha Centauri. The Robinson family are to spearhead a massive emigration of tens of millions of families, making the journey in the prototype *Jupiter 2*. Chosen from two million applicants, John and Maureen and their children Judy, Penny and Will have been judged as 'the unique balance of scientific achievement, emotional stability and pioneer resourcefulness'. Accompanying them is a pilot, Major Don West. He is a public relations exercise; as with the NASA space program of the time, an on-board pilot is not really needed (Wolfe, 1979). Rather, he is there to provide a human face even though their ship will be automatically controlled. He and the colonists will make the whole trip of 98 years in suspended animation. As Don quips, 'if you wake up and find me driving, you'll know you're in big trouble'. He is also the only eligible male for eldest daughter Judy.

Trouble comes with Dr Zachary Smith. In the pay of a foreign government, he sabotages the ship. Unfortunately for him, he gets stuck on board as they take off. Spinning out of control, he wakes Don and the Robinsons, but they are hopelessly lost. In the first six episodes, Smith is a sinister and ruthless secret agent – a classic Cold War villain. Originally cast as a guest star, Harris later camped it up for laughs and a continuing role. However, if we ignore the later Dr Smith and focus on the beginning, he is a dangerous force (really too nasty for a 1960s family adventure). He is not the coward as later portrayed; he easily takes out an armed guard and has no fear of the macho Major West, who he relishes baiting. In sole control of the robot, he is willing to do anything to get them to return to earth. Chillingly, he orders the robot to kill any Robinson he finds alone – although he must make it look an accident. If the family are killed off, Smith reasons, the mission will be a failure and they can turn back. At first, Smith has no qualms about homicide. Later he muses it is a pity that he will have to kill Will because it would be nice to have someone good to play chess with. With Dr Smith, we see a new addition to the explorer narrative. For the first time, one of the party is secretly working against the others and is indeed a cheerfully murderous sociopath. A consummate liar, Smith never breaks cover and the family never realise his mission was to kill them.

In *The Swiss Family Robinson*, the father was the natural leader who looks after and protects his family. Given the period it was made, it might be expected that John Robinson (Guy Williams) would fall naturally into the role of a wise and steady patriarch. However, this was not the case. In the first six episodes, Robinson flounders, constantly putting himself and his family in danger. Under circumstances he cannot control nor cope with, he suffers from a crisis of leadership and a crisis of masculinity.

The problem begins with Alpha Control, who seem to have no concept of a contingency plan. As the intention is that the voyage to Alpha Centauri will be simply accomplished with all in stasis, there is no command structure when the emergency hits. Robinson is only a civilian; he has no skills that are of any use in the crisis. As with many real exploration expeditions, unwise personnel choices are quickly tested. West is unsure what he should do. Maureen Robinson argues with John, particularly when he decides to land to make repairs. Once on the ground, Robinson takes unnecessary risks, trying to prove himself. The younger children – Penny and Will – take no notice of his orders and constantly wander off into danger. Smith watches on smugly; he sees Robinson as a fool who he can easily manipulate.

When they realise that the planet's unusual orbit leads to climatic extremes, Robinson's behaviour becomes more erratic. He decides to abandon the ship and head south. They all nearly perish as a result, whereas Dr Smith stays comfortably put. When an earthquake hits, the rattled Robinson blurts out that this is something he knows about as a trained geologist. Matters come to a head, when Robinson decides they must make a dash back to the ship. West asks for time to check the batteries on their vehicle (the *chariot*). Robinson refuses and they almost come to blows. Predictably, the batteries fail and again they almost perish. Robinson knows it was his fault. West comes to a critical decision. He could take command, but instead chooses to fully support Robinson, covering up the mistake with the batteries. West's decision to slip into the role of loyal lieutenant saves the expedition, for Robinson, despite his failings, is only now accepted as the leader.

Following this rapprochement between West and Robinson, the nature of the story changes. In episode six, for the first time, the narrative shifts to their encounter with a visitor. Their first guest star is Warren Oates playing astronaut Jimmy Habgood, a straightforward human, albeit with a strong Kentucky drawl. He blasted off 15 years earlier, got lost and has been wandering ever since. He has visited 91 planets which, like other explorers over the centuries, he marks with graffiti that he has been there. The Robinson help him with repairs. He could get back to earth and provide directions for a rescue mission. However, Habgood realises that the return home is not what he wants. If he goes back, he will be retired from active service and he is too old to be given another space mission. He intends to go home some time, but not yet; there is still much more of space to explore.

Life of Pi (Yann Martel, 2001)

Canadian author Yann Martel journeys to India to write a novel. Living there is cheap and he hopes he will not be distracted. Unfortunately, he struggles and starts to see that his projected novel will not work. Then he is told a story about a shipwreck. He gets interested and tracks down Pi Patel, an Indian who now lives in Canada. Pi tells his story. However, Martel also interviews a Japanese insurance investigator who suggests the story may be a cover-up. The truth is left open, but the idea of rewriting the narrative to avoid unpleasant aspects, such as the death of individuals, remains. This, it has been suggested, occurred in real exploring expeditions. With Burke and Wills, for example, mystery still remains over the death of Gray. According to Wills' account, he died of starvation and exhaustion, although there is a lingering doubt that an overly severe punishment for stealing food may have been a major factor. In polar exploration, the discovery claims of Frederick Cook and Robert Peary are now claimed to be false (Conefrey & Jordan, 1998). Invention is also a feature in fiction. In *The Admirable Crichton*, after their return, an account is published focusing on the exploits of the gentlemen, with the servants relegated to minor roles. In *Heart of Darkness*, Malone is wracked with guilt after inventing a heroic ending for Kurtz.

Pi is 16 years old. His family run a private zoo in Pondicherry. They decide to sell up and move to Canada. The Patel family travel on a cargo ship across the Pacific, along with the animals that have been sold to North American zoos. When the ship sinks, Pi is pitched into a lifeboat. Along with him are a Zebra, an Orangutan, a Hyena and a Bengal Tiger named Richard Parker (after the hunter who captured him as a cub). The Hyena kills and eats the Zebra and Orangutan. In turn, the Tiger kills the Hyena. Now there are just Pi and a seasick Tiger. Pi builds a makeshift raft, which he attaches by rope to the lifeboat.

How to survive in a confined space with a dangerous animal? Pi formulates a plan:

> I had to tame him. It was at this moment that I realized this necessity. It was not a question of him or me, but of him *and* me. We were, literally and figuratively, in the same boat. We would live – or we would die – together ... But there's more to it. I will come clean. I will tell you a secret: part of me was glad about Richard Parker. A part of me did not want Richard Parker to die at all, because if he died I would be left alone with despair, a foe even more formidable than a tiger ... He pushed me to go on living ... It's the plain truth: without Richard Parker, I wouldn't be alive today to tell you my story. (Martell, 2001: 164)

Having grown up in a zoo, Pi understands theories of animal hierarchies and has over the years heard stories of how lion tamers use this to their advantage. Using food, a whistle and the ability to rock the boat (making the tiger seasick), Pi establishes himself as the dominant entity. Fishing and tiger training become the daily routine, both necessary for survival:

> I wonder if those who hear this story will understand that my behaviour was not an act of insanity or a covert suicide attempt, but a simple necessity. Either I tamed him, made him see who was Number One and who was Number Two – or I died the day I wanted to climb aboard the lifeboat during rough weather and he objected. (Martell, 2001: 206)

For 227 days, these two unlikely castaways float across the Pacific. When a container ship comes in sight, Pi thinks he is saved. Yet the crew do not notice him and he is almost run over. When he sails through a patch of rubbish, he retrieves a bottle. Desperate, he puts a message in it and throws it overboard. It reads, 'Am in a lifeboat. Pi Patel my name. Have some food, some water, but Bengal tiger a serious problem' (Martell, 2001: 238). One wonders what anyone would make of that. Eventually they make landfall in Mexico. The tiger charges off into the jungle. Pi has no proof of his story.

The Lost Explorers

These narratives of lost adventurers provide an extra dimension to the explorer myth. Of course, all explorers are lost, for on their expedition they are cut off from home, often unsure of where they are and how they will get back. Thor Heyerdahl and his Kon-Tiki Expedition were a good example of this as they drifted across the Pacific Ocean, without any steering and at the mercy of currents. Being lost adds danger and uncertainty, testing the heroic explorer. Most of all, being lost highlights the importance of being found or rescued in order for there to be the return narrative.

8   Re-enactments

Not all explorers aim to re-enact past journeys. Isabella Bird, for example, 'was not interested in following the track of Cortés, of Hannibal, or of Alexander the Great' (Boorstin, 1960: xviii). Many, however, do, and their fascination with walking in the footsteps of travellers shows no sign of abating. Indeed, the travel writer Rolf Potts describes it somewhat cynically as 'the presumed formula of what travel writers were supposed to do ... the travel-literature equivalent of cover music – as common (and marketable) as Whitney Houston crooning Dolly Parton tunes' (Potts, 2008: xiii). It is also a ubiquitous thematic structure for many television travel series, often with a celebrity host, such as Bernard Levin's *Hannibal's Footsteps* (1985), Michael Palin's *Around the World in Eighty Days* (1989), David Suchet's *In the Footsteps of St Paul* (2012) and Michael Wood's *In the Footsteps of Alexander the Great* (1997). Potts' critique of this device is that it prevents the traveller from meandering through a place in a way that is non-linear, contradictory, flexible and non-thematic. While this might be true in some instances, this mode of travelling provides many explorer travellers with highly satisfying experiences that clearly meet a variety of needs, which warrant a deeper examination. It also has a long heritage, stretching back far beyond the modern media age. In fact, it can be argued that all travel is a form of *ritual repetition*, where the individual treads a well-worn path (Akerman, 1993; Edensor, 2000; Pearce, 1979; Seaton, 2002) or gazes on what has already been seen by others (Urry, 2002).

In this chapter, we examine the modern trend for explorer travellers, often elite adventurers, to construct expeditions recreating the travels of past explorers. These often have a strong media focus, partly to gain sponsors, and are successful in capturing extensive public interest. Many are couched in terms of solving mysteries and/or reassessing the reputations of past explorers. For some, there are spiritual overtones, as they pay homage to their heroes of the past. These experiences may also involve role play, as the traveller imagines themselves *as* the explorer or begins to feel closer to them and/or to try to understand what made them undertake these extraordinary journeys.

A Rationale for Dangerous Travel?

As Potts (2008) alleges, one potential reason for re-enactment of past journeys is that it gives travel a purpose or *gravitas*, which masks the indulgence of expensive and highly risky travel. Couching an expedition as a re-enactment is thus similar to attaching pro-social motivations to it – it makes the traveller feel less guilty about what they are doing. Anker and Roberts (1999: 29) refer to the distrust that was directed at the expedition to try to recover Mallory and Irvine's bodies from Everest:

> Most observers, however, viewed the expedition as something of a boondoggle – one more stratagem, like campaigns to raise money for medical research or to clean up other expeditions' trash, to finance an expensive outing onto the world's highest mountain.

Peter Hillary linked his polar expedition with those of a bygone era, ostensibly to satisfy sponsors. He labelled his journey to the South Pole in 1998 'a minor footnote' to the story of Scott and saw his commitment to sponsors as continuing a tradition and as a necessary evil: 'That's the way it has always been done. Adventurers and explorers have been promising the world to their patrons for centuries ... [The publicity] was too much. It was a little out of control. It was the only deal going. And I think the Captain [Scott] would have taken it too' (Hillary & Elder, 2003: 26, 27). His reference to Scott does, however, suggest that he sees himself as a modern heir to past explorers. Rather than merely a shrewd excuse to raise money, his allusions to Scott's journey may in fact represent *metempsychotic* behaviour.

A re-enactment also provides a framework or structure to a journey, which might give the impression that the travel is less dangerous than it otherwise might be. After all, it has been done before! Yet some places have become more hazardous in recent times, like Iraq, as is illustrated in the following narrative by Eames. This arguably gives the traveller more kudos or prestige than the person in whose path they are treading. Thus in honouring a traveller of the past, they are also heightening their own status and reputation as a traveller.

The 8.55 to Baghdad (Andrew Eames, 2004)

Andrew Eames, a British travel writer, is a fan of re-enactments. His recent book, *Blue River, Black Sea* (2010), documents a sojourn along the Danube, in the footsteps of Patrick Leigh Fermor. Before that, he recreated a train journey made by Agatha Christie in 1928 to Iraq, fresh from the breakdown of her first marriage, suffocated by life in a suburban golf estate, and looking to find out what she was made of. As the story goes, she found love, in the form of archaeologist Max Mallowan, and discovered a new life among the excavations at Ninevah and Nimrud. Eames hears about Agatha's connection with the Middle East on a trip to Aleppo, where he meets the mother of the owner of the Baron Hotel, who had encountered the Mallowans on their stays. Eames is intrigued: 'What was the crime writer doing in such an out of the way place?' (Eames, 2004: 8). He gives his ignorance of her travel history as the reason for his decision to retrace her steps: 'For me that was a beginning, not of a whodunit, but a whydunit and how' (Eames, 2004: 11).

This is a somewhat spurious rationale for a highly dangerous journey in the lead-up to the 2003 invasion of Iraq. The Christie re-enactment perhaps gives the journey a serious *raison-d'être* which it would otherwise lack, being merely a foolhardy jaunt to what would become a war zone, where even writing notes and taking photographs could put the traveller in danger (similar to Danziger, 1988; see Chapter 9). Eames is not alone for the Iraqi leg of his trip, having joined a tour group whom he paints as eccentric misfits, reminiscent of an ensemble out of a Christie novel, including fussy old ladies like Miss Marple and an old fogey like a young Poirot. Eames sees himself as an outsider to the group and is quick to mock them, like others he encounters on the road. In this way, he is the quintessential superior traveller. He has no patience for human foibles, and a sharp tongue when describing physical appearance, even when it does not seem pertinent to the narrative. Even Eames' departure from Sunningdale Station, where Agatha lived with her first husband Archie, is an occasion to sneer at the moneyed inhabitants.

Eames engages us when he becomes more introspective. He is finally honest about the reasons for his trip when he refers to his need to escape from his everyday life. He notes that domestic discussions 'in the middle period of life' tend to 'get top billing, to the exclusion of any more fundamental impulse to realize one's own destiny' (Eames, 2004: 19). His journey, echoing Agatha's, is a chance to regain control over one's life – 'some sort of re-assessment from a distance was overdue. I was hoping that the process of travel itself would clear the palate, like a Calvados between courses' (Eames, 2004: 19). Eames also acknowledges the attraction of risk and challenge – 'there are few journeys which are far more complex and difficult than they were seventy-five years ago, but to travel from London to Baghdad, by train, is one of them' (Eames, 2004: 20). While he likens himself to Christie, in their love of trains and travel, and their lack of pretence that they were 'creating great literature' (Eames, 2004: 48 – another barb or disingenousness?), he also paints himself as not merely travelling in Agatha's footsteps, but outstripping her: 'Her adventure had been undertaken at a moment of major personal change; mine was beginning at a moment that could change the world' (Eames, 2004: 20). He ends the book by reiterating that it is this additional layer which makes his journey more important than the one he initially envisaged:

> My interest in Agatha and her crisp pieces of fiction had finally been overhauled by a far bigger story, a whodunit where the world might have been still unsure of the actual nature of the crime but the punishment is already on its way. (Eames, 2004: 377)

Eames draws out the ephemerality of life for the reader, by documenting in great detail the changes to trains, hotels and the houses where Agatha lived during her expeditions to the Middle East. This is where a re-enactment can shed light on societal transformation, as well as mere changes to bricks and mortar. He traces the lives of those who survived communism and civil war in the Balkans, for example, and is at his best when he stops judging and starts listening to stories that people seem all too willing to tell to a stranger in their midst. The crumbling of the house where the expedition to Ur was based, on Agatha's first trip to Iraq, symbolises the overthrow of regimes and the decay of societies, ironic in a landscape where the past is constantly being brought back to life: 'The 6,000-year-old Ziggurat may have been still standing but every last trace of a 100-year-old dig house had been wiped away' (Eames, 2004: 376).

He has a moment of epiphany on a bus between the Iraqi border and Baghdad, where he muses about Agatha Christie's book *Absent in the Spring*, which she wrote under an alias, Mary Westmacott. It is about a woman who is isolated by floodwater on a journey home from Baghdad, and is forced to strip away the veneer she has carefully built up: 'The result was a real crisis of identity brought on by silence, loneliness, the desert sun and middle age' (Eames, 2004: 288). Eames observes that travel such as he is undertaking can be confronting, and that we need others towards whom we can relate, to be the best form of ourselves. Through his re-enactment, he gets closer to Christie than he had perhaps anticipated.

Solving a Mystery or Conundrum

The most well-known re-enactment aimed at solving a mystery is Thor Heyerdahl's *Kon Tiki* expedition (Laing & Crouch, 2011), in which the anthropologist sailed on a wooden raft from South America to Polynesia.

This was a scientific experiment, in order to prove that this was how Polynesia was originally populated. His journey showed that his premise was possible, but unfortunately more recent research, including DNA and archaeological evidence (Hunt & Lipo, 2011; Hurles *et al.*, 2003), does not support his theory about the origins of the Polynesian people, and suggests that they came from Asia. Nevertheless, the scale of the re-enactment and the risks that Heyerdahl undertook on this epic voyage are legendary. The book he wrote of his adventure was translated into 70 languages and sold 50 million copies, while his film won an Academy Award for the best documentary feature. The Kon-Tiki Museum in Oslo celebrates his achievements.

His modern heir is the historian Tim Severin, who is equally known for his frequent re-enactments. He saw his recreated sea voyage of Saint Brendan as 'a detective story', which needed to be solved, in order to determine whether the Irish saint could have crossed the Atlantic:

> I had the clues before me in the text of the *Navigatio*. One by one they might lead me toward a solution, providing I could find out how to follow them? But how? Again, the obvious answer was with a boat exactly like the one Saint Brendan had used. Such a boat would take me to inspect the places along the Stepping Stone Route that might conform to the places recorded in the *Navigatio*. At the same time it would also show whether such a boat could survive an Atlantic voyage. But what exactly did the *Navigatio* mean by a boat made of oxhides stretched on a wooden frame? Could such a vessel make an Atlantic crossing? (Severin, 1978: 15)

The Mallory and Irvine mystery has similarly beguiled many explorer travellers, who wanted to know if the pair were the first to climb Mt Everest. For this reason, Conrad Anker wanted to 'try to free-climb the Second Step. That, for me, was the crucial test of the likelihood that Mallory and Irvine could have made the summit' (Anker & Roberts, 1999: 71). He could then see how difficult it would have been for the climbers in 1924, and thus assess the veracity of the claims that they died on the way *down* from the summit, not the way *up*. Anker later conceded that, while he wanted to believe the former, he thought that the correct interpretation of the evidence gained from his expedition was that they did not reach the summit. Climbing the Second Step with the technical gear and clothes that they used in 1924 'would have been an utterly terrifying proposition' (Anker & Roberts, 1999: 165). This was the clincher, but there were other clues as well, such as the fact that Mallory's body showed no sign of frostbite, which would have been present if they had spent a night in camp after reaching the summit.

Another historical conundrum concerns whether either Frederick Cook or Robert Peary was the first person to reach the North Pole and indeed whether either of them made it there at all. Paul Landry, who guided Patrick Woodhead to the South Pole, was fascinated by the Peary and Cook tale. He 'spent years reading the various arguments and counter-arguments. One day, he put down his pen, returned the books to the library and went to see for himself' (Woodhead, 2003: 89). Landry arrived at the Pole 'only four days behind Peary himself' and his expedition 'certainly proved [Peary] could have made it' notes Woodhead (2003: 90), although the jury is still out on this feat (Conefrey & Jordan, 1998).

Solving a mystery might be undesirable, destroying the attraction of the unknown. Keith spoke with us about his attempts to find Lasseter's Reef. In 1929 Harold Lasseter announced that he had discovered a fabulous gold reef in a remote part of Central Australia. His timing was perfect; the onset of the Great Depression made the public hungry for such an El Dorado. With government assistance, he put together an expedition to return to the reef. However, once in the Outback things began to unravel. Either Lasseter was a fraud, could not find his way back, or was paranoid and laying a false trail. Separated from the others, he perished. While his body was recovered, the reef was never located (McGowan, 2006). When we interviewed Keith, he explained:

> A few times I've been out to look for Lasseter's lost reef. I don't want to find it. I said 'No-one must ever find it', because we want to keep looking for it forever. Young people can organise their expedition and look for Lasseter's Reef. If some fool finds it, you can't do that anymore!

Some re-enactments involve the search for a missing explorer. These include Stanley's journey to find Livingstone, various attempts to find Mungo Park, Ludwig Leichhardt and Percy Fawcett, and the expedition by Roberts and Anker to discover the bodies of Mallory and Irvine on Everest. Finding out the fate of these individuals is important – it honours their sacrifice and brings some degree of closure to their relatives. It may also become a type of pilgrimage. Mallory's body takes on the guise of the Holy Grail or a medieval relic and Anker and Roberts (1999: 39) concede that Mallory and Irvine's disappearance has been elevated 'to the realm of the mythic' due to historical conjecture as to whether they reached the summit before they died. Another climber, Thom Pollard, who discovered Mallory's body during the recovery expedition, also expresses his deep respect for a fallen hero: 'I kept thinking, this is one of the greatest figures in the history of mountaineering. I got down on my knees and asked for guidance.

I prayed, in essence, for guidance not to desecrate the site' (quoted in Anker & Roberts, 1999: 139).

Illustrative of these attempts to discover lost explorers, we turn now to a narrative describing a journalist's search for Fawcett and his city of Z.

The Lost City of Z (David Grann, 2009)

David Grann is a reporter for the *New Yorker*. It's the dream job for a writer – one of the most prestigious publications in the world and, as he says himself on the inside back cover of *The Lost City of Z*, he covers 'everything from the presidential campaign to the hunt for the giant squid'. It wasn't enough. Grann needed an adventure. He is blunt about his qualifications for the task, using the self-deprecating tone beloved of writers of explorer tales (Laing & Frost, 2012; Murray, 2008; Thompson, 2011):

> Let me be clear: I am not an explorer or an adventurer. I don't climb mountains and hunt. I don't even like to camp. I stand less than five feet nine inches tall and am nearly forty years old, with a blossoming waistline and thinning black hair. I suffer from keratoconus – a degenerative eye condition that makes it hard for me to see at night. I have a terrible sense of direction and tend to forget where I am on the subway and miss my stop in Brooklyn ... Given a choice each day between climbing the two flights of stairs to my apartment and riding the elevator, I invariably take the elevator. (Grann, 2009: 25)

But then he cuts to the chase. What is it that entices him away from his 'newspapers, take-out food, sports highlights (recorded on TiVo) and the air-conditioning on high' (Grann, 2009: 25)? It is an allure that has been with him since childhood, when he first read 'mystery and adventure tales, ones that had what Rider Haggard called "the grip"'.

His literary inspiration to travel, with its genesis in his earliest years, is reminiscent of a number of adventurers. In our interview with Rod, he observed how he:

> Loved reading the *Biggles* books and *Tintin*, which had adventurous elements in them of course, and [to a] 10, 12, 13 year old, that's kind of what grabs your attention because you like comics and they're always idyllic adventures because they always come out good in the long-run. And they have little epics along the way as you are supposed to and I felt like I was living those people's lives. Then I just had the desire to make it reality, I suppose.

Grann is also influenced by real-life adventurers, such as his grandfather, a freelance *National Geographic* photographer. This again resonates in our interviews. Robert, for example, recounted the importance of childhood influences, such as his childhood heroes, on his desire to sail and take part in round-the-world sailing adventures:

> It's amazing the influences you pick up when you're a kid and it's very important as you develop further on, so your formative years still have relevance to what you do as you get older and you get money and can do things. And I suppose sailing became an issue in that I was influenced by my hero, as a boy. It was a guy called Robin Knox-Johnston and he was the first guy to sail single-handed around the world non-stop in 1968. I sort of followed it in the papers and I was completely blown away by this guy.

Drawn to adventurous assignments, Grann understands the fine line between adventure and obsession, using the metaphor of cancer, acknowledging how 'ordinary people … get some germ of an idea in their heads that metastasizes until it consumes them' (Grann, 2009: 27). His profession requires these qualities to a degree, when he hunts down leads for a story and seeks the often hidden truth in a morass of red herrings and rabbit holes. His discovery of Fawcett's doomed 1925 expedition came while he was researching another story on Conan Doyle and realised that the true-life adventure had inspired Conan Doyle's *The Lost World*. Grann was intrigued by Fawcett's premise – that there had been an advanced and cultured civilisation living in the heart of the Amazon jungle, the so-called Lost City of Z – and he wanted to investigate this mystery, as well as finding out how and when Fawcett died, along with his son and his son's schoolfriend.

Grann was not the first to seek to recreate Fawcett's footsteps. Many had tried to solve the enigma of his disappearance, including Fawcett's other son Brian, and some had died trying. Was he killed by Indians, wild animals or disease? What about the two younger men – Fawcett's son and his schoolfriend? Grann is fully aware of the danger of what he is attempting, but is caught up in the quest and cannot rest until he succeeds. Grann encounters dead ends and realises he has a story, but not the one he intended. He still is no wiser as to the fate of Fawcett, nor does he have any tangible evidence that the city of Z ever existed. Should he just go home?

The twist to the tale is that at the start of his quest, he has been invited by an archaeologist from the University of Florida, Michael Heckenberger, to join him in a village in the Xingu, which is the region where Fawcett was last seen. Thinking he has nothing to lose, Grann meets up with Heckenberger before he leaves the Amazon, and paradoxically finds what he has been seek-ing all along. He is taken to see a series of moats, then a wall, and shown shards of broken pottery and roads: 'Heckenberger said that we were stand-ing in the middle of a vast ancient settlement' (Grann, 2009: 269). He had discovered 20 settlements connected by roads, of pre-Columbian origin. Fawcett was right. Here was Z.

Personal Challenge

Some re-enactments aim to complete what was started long ago. They might be a personal quest of a family member. John Harlin III set out to climb the north face of the Eiger where his father perished in 1966, plunging 4000 feet due to a broken rope (Harlin, 2007). His life up to that point involved climbing, but he often questioned why he needed to do it: 'I think too many people are dying in the mountains, are they worth it?' (Harlin, 2007: 197). Like the Mallory and Irvine rescue, this journey is a pilgrimage, 'not simply a route' (Harlin, 2007: 224). Harlin Jr. writes of looking at the summit of the Eiger: 'I felt like a religious devotee at a sacred shrine' (Harlin, 2007: 203). It is also a test of his manhood – a *rite of passage* (Turner & Turner, 1978; Wilson, 2000). He wonders whether people are comparing him (unfa-vourably) with his father and how he would fare up there. It preys on his mind: 'It seemed I couldn't get away from the Eiger. Not that I was trying to' (Harlin, 2007: 207). Eventually he realises that he cannot rest and would always feel a failure in his father's eyes if he did not attempt the climb. Even his guilt at leaving his daughter potentially fatherless, as he was, does not stop him. Climbing the Eiger will give him closure – 'to climb out from this specter so we can get on with our lives' (Harlin, 2007: 261).

The Matrix Shackleton Centenary expedition in 2009, discussed later in this chapter, involved a team of three men, Henry Worsley, Will Gow and Henry Adams, all related to one of the men on Shackleton's journey, which failed to reach the South Pole in 1909. The modern adventurers reached the point where Shackleton's expeditionary team turned back, but then contin-ued on to the South Pole. It was seen as a way to honour Shackleton and his bravery, including his decision not to endanger his men by pushing on. As Worsley noted on the official expedition website: 'Setting out to close the last 97 miles will be a huge prize and I hope will prove to be a fitting legacy to the original expedition in the centenary year' (Matrix Shackleton Centenary Expedition, 2013). Once he reaches the Pole, he pulls out Shackleton's com-pass, which he had carried with him throughout the whole journey: 'Irrevocably, a part of him had finally reached 90° South' (Worsley, 2011: 231).

These challenges are not necessarily about matching the exploits of the past. The mere attempt to do so may be reward enough. Jeffrey Tayler re-enacts the myths bound up in Thesiger's masterpiece *Arabian Sands* (1959) and is initially disappointed at not being able to reach his literary hero's exacting standards during his expedition through the Sahara:

> I was falling short of Thesiger, who disparaged the amenities of civiliza-tion and 'learnt the satisfaction which comes from hardship and the plea-sure which springs from abstinence,' and who, among the Bedouin, found 'comradeship in a hostile world'. (Tayler, 2003: 174)

As he notes, however, at the end of his journey: 'we were not Thesiger and his Bedouin, but lesser mortals who had walked a long way, suffered a lot, and exhausted themselves. Still, we had done what we had set out to do' (Tayler, 2003: 234).

One of our interviewees, Charlie, who recreated a journey of Australian explorers, was intrigued by the notion of succeeding where his historical counterparts had failed: 'I just thought the connection between two explor-ers and what they couldn't do back then, and what we could do now, yeah, the whole thing was fascinating'. This is part of the challenge and arguably bestows upon the explorer traveller a sense of prestige or status for achieving what their forebears couldn't. His comment echoes adventurer John Hare's motivation for recreating the explorer Hanns Vischer's Saharan expedition. Vischer took the trip in reverse in 1906, but was prevented by his employer from going back the other way. Hare wanted to 'make the journey Vischer had been unable to undertake' (Hare, 2002: 62). Roger Mear was similarly intrigued by the challenge of retracing Scott's footsteps and getting to the Pole: 'Two men, hauling sledges on the longest white walk in history' (Mear & Swan, 1987: 29).

Resurrect Reputations

This motivation behind a re-enactment is similar to that underpinning the completion of a quest. It is all about the current generation respecting and acknowledging the deeds of the past. A forgotten explorer, like a lost explorer, needs to be rediscovered and paid tribute to, as a hero. This is the myth-making element behind a re-enactment (Laing & Crouch, 2011). We spoke with Geoff about how he wanted to bring to light the story of the explorer Sir John Ross: 'It's got to be told, it's been forgotten. So yeah, that is a big motivation for me'. Kieran Kelly (2000: 254) was also keen to see the explorer Augustus Gregory recognised for his achievements – 'this will be my reward'. He enjoyed telling schoolchildren about his expedition in the footsteps of Gregory and was pleased that one school decided to create a course on Gregory after hearing his presentation (Kelly, 2000).

There may also be a desire to *debunk* myth which has grown around some explorer narratives. Robert Swan saw Scott's fatal journey to the South Pole in this light, with the polar explorer often criticised in more recent times for his perceived blunders, particularly the use of ponies instead of dogs in his expeditions. His trekking partner, Roger Mear, explains why the re-enactment was so important to Swan:

> We are a more cynical nation now and our heroes have been toppled from their pedestals. They have been denounced because they have been revealed as men with weaknesses and failings. Scott fell to the biased pen of Roland Huntford in his book *Scott and Amundsen*. Yet the truth lies somewhere between the two extremes. Robert's dream was not simply to walk to the South Pole but to rediscover Scott and bring alive the histories of those men who so inspired him. (Mear & Swan, 1987: 27–28)

Furthering Knowledge

Many re-enactments are couched in terms of their contribution to academic scholarship and science. Historian Tim Severin, in deciding to undertake the Brendan Voyage, in the footsteps of the Irish monk, notes 'First, I had to satisfy myself that the scholarship behind the project was sound ... To warrant such risk and effort the endeavor had to produce worthwhile results' (Severin, 1978: 10–11). In some cases, this might be a way of covering the journey with a veneer of worthiness; akin to pro-social motivations like fundraising for charity or inspiring a younger generation (Laing, 2006). The Shackleton Centenary Expedition took part in an experiment by collecting ice samples for Hull University (Worsley, 2011). There were concerns that this might distract them from their main aim, but they were keen to assist a geographer who had helped to promote their expedition to potential backers. Tim Mackintosh-Smith (2001: 12) also refers to a quasi-scientific rationale for his re-enactment of Ibn Battutah's travels, which he labels 'inverse archaeology ... I would start with a life – IB's – and go off in search of its memorabilia'.

Occasionally, universities will fund expeditions as a means of conducting academic research, as illustrated by then undergraduate William Dalrymple's journey in the footsteps of Marco Polo.

In Xanadu: A Quest (William Dalrymple, 1989)

William Dalrymple is not the first traveller to attempt to recreate Marco Polo's journey and write about it. Clarence Dalrymple Bruce wrote *In the Footsteps of Marco Polo: Being the Account of a Journey Overland from Simla to Pekin* (1907) and, in the same year as Dalrymple's account, Jin Bohong wrote *In the Footsteps of Marco Polo* (1989). Dalrymple, however, had the chutzpah to get his prestigious university (Cambridge) to fund his expedition, and set off in his holidays, keen to return home before term began. Like Eric Newby, he utilises a trope of the bumbling Englishman at times, contrasting his naiveté with the assertive streetwise style of his two female companions – Laura and later Louisa.

His book does not directly engage with the arguments put forward by the likes of Wood (1996), who posits that Polo never reached China. Instead he wants to 'put flesh and blood on the dry bones of the world of Polo's *The Travels*' (Dalrymple, 1989: 43), rather like the writings of Usamah inb-Munquid, which he reads on the way and finds riveting in its depiction of the Middle East in medieval times. Dalrymple often seeks to verify details of Polo's account, and refers to finding a silk loom 'that proves Polo's accuracy in all things mercantile' (Dalrymple, 1989: 127). In other instances, he exclaims at the differences between what he sees and the descriptions in Polo's book, like dry landscape that was once rich pasture, and the failure to find the bodies of the Three Wise Men at Saveh. The 1929 edition of *The Travels* is referred to as the 'expedition bible' (Dalrymple, 1989: 84) and is consulted at each stop to see where the team has reached and how far they have still to go.

Dalrymple is a romantic at times, with references to places being 'straight out of *Sheherazade*' (Dalrymple, 1989: 49), but is also brutally honest about the challenges of his journey, including indifferent hygiene, the lack of or unfathomable timetables and the often frustrating border crossings. As he notes at one stage: 'There are moments in all long journeys when the whole business of travelling seems utterly futile. One feels homesick, tired and above all bored. Nothing pleases. Everything palls' (Dalrymple, 1989: 220).

Like Richard Burton, he employs a costume at times to maintain a disguise, in this case to help him blend in and avoid border controls. There is also a sense of Dalrymple enjoying the feeling of being an explorer, if not actually taking part in role play. He compares himself to Burton in feeling depressed at finally achieving his goal, and notes that he might share the 'honour' with Alexander the Great of being the only two outsiders to 'be presented with a goat's leg by the Gujars' (Dalrymple, 1989: 212). There are also numerous comparisons with Marco Polo, including similar illnesses, and Dalrymple's concern after leaving Xanadu 'that nobody would believe our tale, just as they had refused to believe Marco Polo himself' (Dalrymple, 1989: 301). Ultimately, however, Dalrymple is keen to have the expedition classed as a contribution to knowledge. The Cambridge name opens doors and, while Dalrymple laughs at the hyperbole of the official letter of introduction his lecturer wrote for him ('If it is to be believed, any obstacle to our expedition could well prove a major blow to the study of the Orient as we know it'; Dalrymple, 1989: 182), the fact that he reprints the letter in full suggests a certain pride in its contents, including its reiteration of the academic merit of the expedition. Dalrymple also later notes that the fragments of tile he gathered from Xanadu 'were later dated by the Fitzwilliam Museum as thirteenth-century Mongol' (Dalrymple, 1989: 301). As his journey finishes, his mind turns to university finals, and no doubt the reception of his journey within the hallowed halls of Cambridge. The book ends with evidence of careful documentation, as befits a serious academic – 'I got out a sharp pencil, opened a blank page and began to scribble' (Dalrymple, 1989: 302).

Authenticity: The Explorer as a Role Model

A re-enactment might be based on the fascination of seeing what an explorer saw in the past, and experiencing what they did; a form of authenticity based on undertaking real, visceral experiences. Tim Mackintosh-Smith (2001: 11) reads about the medieval pilgrimage of Ibn Battutah and is captivated by his story. It becomes an 'itch' for Mackintosh-Smith and he realises that virtual travel will never be enough – he needs to depart on his own version of Battutah's journey. For Kieran Kelly, seeing Australia through the eyes of an explorer was partly the motivation behind his recreated expeditions through the desert. An example is his journey with Andrew Harper in the footsteps of explorers Gregory and Stuart, hoping to complete their journey across the Tanami desert: 'I was excited by the prospect of seeing central Australia, seeing what the little Scotsman [Stuart] had seen, walking where he had ridden and finding out what sort of person he had really been' (Kelly, 2003: 53).This is close to role play, discussed below, but also involves the explorer acting as a *role model* (Laing & Crouch, 2011; Seaton, 2002) for the modern journey.

The re-enactor is thus looking for an *authentic* experience, in terms of the degree of risk and challenge, the existence or extent of technology, support, guides and rescue, the type of motivations involved and the carrying on of traditions from years gone by. It also gives them a form of *existential authenticity* – 'seeking a space in which they can construct self-identity and be their true selves' (Laing & Crouch, 2011: 1528). Kieran Kelly writes that his re-enactment of Gregory's 1885–1886 expedition through the Northern Territory of Australia became a journey of 'personal discovery':

> Like Huck Finn travelling down the Mississippi with Jim, I learned a lot about myself. Gregory was a better horseman than me, a better navigator and a more capable leader of men. Designed to be an enlightening experience, the North Australian Expedition 1999 became a humbling one. (Kelly, 2000: 254)

While most explorer travellers recognise that they are not carrying out these journeys in *exactly* the same way as their heroes did, this does not make these experiences staged or fake. There are still high risks involved and no certainty of success (Laing & Crouch, 2011). Roger Mear points out that to a degree he was prepared to overlook or tolerate the artificiality of his re-enactment of Scott's expedition, if the challenges were great enough:

> Of course, it could never be a two-way journey on foot as Scott's had been, for no one could drag the weight of five months' food, but the thought of being out there, in such awesome isolation, with just a pair of skis, a sledge and a tent, made the artificiality of a one-way journey acceptable. To a mountaineer like myself it offered all the mystery and challenge of an unclimbed ridge on Everest or K2 or the perpendicular granite walls of Yosemite. (Mear & Swan, 1987: 29)

There can be disputes over what constitutes an authentic experience, and whether things have been done in a *pure* way. Bettina Selby encountered criticism of her journey on these grounds, but was pragmatic about what she had achieved:

> Many people told me that I had not reached the ultimate source of the Nile. In the hills of Barundi, they said, there was a pipe trickling out water and near it a German had built a small pyramid to commemorate all the Nile explorers, which proved it must be the true source. Others laid claim to different sites. I didn't argue with any of them, for I had not been on that kind of journey and I was not a geographer. The Nile had a multitude of sources, as befits so great a river. (Selby, 1988: 231)

Pilot Peter Norvill, in his re-creation of Sir Charles Kingsford-Smith's flight, saw reliance on modern technology as taking away some of the challenge of his endeavour. He was pleased when his equipment failed, placing him on the same level as Kingsford-Smith:

> I was about 20 minutes out [of Hawaii] when I experienced the instrument pilot's nightmare. The vacuum pump failed ... To change them would have meant turning back to Hilo and by the time I had done so and fitted the new pump it would have been too late to attempt to reach Christmas before dark, which would have meant another night in Hilo and the loss of another day's progress ... This effectively reduced me to the same standard of instrumentation that Smithy had in the 'Southern Cross' 60 years ago when he was flying this route to Brisbane. (Norvill, 1988: 131)

For this reason, Mear and Swan did not take a radio with them on their re-enactment of Scott's journey. Swan viewed walking across Antarctica without radios as 'an opportunity for truth' (Mear & Swan, 1987: 84), while Mear saw it as a way to demonstrate some of the extraordinary dedication shown by the explorers in Scott's day – 'to play the game by the old rules' (Mear & Swan, 1987: 234).

Perceived authenticity can, however, be shattered by a chance encounter with the outside world. Patrick Woodhead found a footprint during his Polar trek and observes: 'In one way it felt a little annoying that someone had evidently trodden here before us' (Woodhead, 2003: 205). Roger Mear was similarly upset to meet another traveller during his re-enactment, arguing that it had 'destroyed the mystery and doubt that had hung over the remaining 480 miles' (Mear & Swan, 1987: 180–181).

Role Play

The desire of individuals to role play and act out explorer narratives has been examined in a number of studies, notably Gyimóthy and Mykletun (2004). Hall (2002: 295) cites the ecologist Aldo Leopold, who argues that 'the opportunity to relive or imagine the experiences of pioneers or the "frontier" that formed national culture ... was an essential component of the value of wilderness and wilderness recreation'. It is also a key hallmark of re-enactments, where the traveller sometimes takes on the persona of their hero and imagines themselves in their place, or conversely envisions themselves alongside them (Frost & Laing, 2013).

Bettina Selby travels through Africa, in the footsteps of Victorian traveller and novelist Amelia B. Edwards, and has created a colonial fantasy around her modern journey. In Aswan, Selby 'plays' at being a lady of those colonial times (Selby, 1988: 69) and imagines the waiters as her personal staff: 'All my leisure moments I spent on the terrace of the Old Cataract [Hotel], drinking turkish coffee brought by Nubian servants in red fezzes.' Selby's fantasy also encompasses following in the footsteps of the 'heroic' colonial explorers, whom she later refers to as 'my Victorian forerunners':

> After Khartoum I couldn't hope to be near a postbox or a telephone again for some considerable time, nor was I likely to meet a returning traveller who could take a message for me. Like the Victorian explorers before me, I anticipated being out of touch for a long while – always assuming that I did succeed in getting through and wasn't bundled straight back to Khartoum again – a fate that also hung over the head of my hero Samuel Baker when he and Florence had set off on this stretch at much the same time of year in 1862. (Selby, 1988: 168)

Selby felt a kinship with explorers of the past and saw herself as sharing characteristics or qualities which inspired her recreation of a historical journey, in this case travelling to the source of the Nile, in tribute to her explorer heroes or heroines: 'I went to pay my respects to Speke's memory at the place where the Nile really begins, where the waters of Lake Victoria start their exit through the Ripon Falls' (Selby, 1988: 231).

Revisiting the Past: Scott, Shackleton and Mawson

Polar re-enactments often combine all the elements we have discussed in this chapter, and we felt that their popularity in recent times warranted separate examination to conclude this chapter. There is an obsessional quality to these experiences, more so than some of the other forms of explorer travel we have considered. Roger Mear, for example, became increasingly alarmed at his teammate Robert Swan's fanatical fascination with the legend of Captain Scott and relentless drive towards success at all costs. This obsession led Swan to seek to emulate his hero, almost to the point of death, and perhaps, in so doing, achieve the same legendary status as Scott and his tragic compatriots:

> 'There is a side of me' Robert [Swan] admitted softly, 'that wouldn't mind dying out there. In fact, I would quite like it in a way'. (Mear & Swan, 1987: 135)

Authenticity is also an integral part of these journeys, underpinning what is done and the reasons why. Mear and Swan (1987), for example, note that their brochure 'proudly proclaimed, like the opening lines of an episode of *Star Trek*':

> An Expedition whose intention is to retrace Captain Robert Scott's foot-steps to the South Pole and to restore the feelings of adventure, isolation and commitment that have been lost through the employment of the paraphernalia of modern times. Without recourse to depots, dogs, air support or outside assistance of any kind, two men alone will manhaul the 883 mile journey to the Pole.

However, the circumstances are different between the modern journeys and their historic precedents. Patrick Woodhead, for example, does not place himself in the same category as the heroic explorers of yesterday:

> Technological developments in air travel and computers have compacted everything, diminishing the last great wildernesses. Suddenly, planet Earth is a much smaller place. For good or bad, this fact enables people like me, who are fragile and unadventurous in comparison, to follow in the footsteps of the great explorers and reach for the same goals. We are the product of that advance and piggyback off it to our advantage. (Woodhead, 2003: 122)

Not all polar re-enactments involve derring-do. The Oamaru Scott 100 Centenary was a way for the small New Zealand town, best known for its Victorian heritage and steampunk subculture (Frost & Laing, 2014), to put its name on the map. In 1913 the *Terra Nova* arrived in Oamaru's harbour and two crewmen rowed ashore and sent a telegram, advising their London backers of the death of Captain Scott and his four men. At 6am on 10 February 2013, this event was re-enacted by locals and actors, playing the crew, the night watchman and the harbourmaster. A local actress, playing Scott's wife Kathleen, read out extracts from her memoirs (Bruce, 2013). We were fortu-itously in Oamaru for this commemoration, and attended an event outside the old Post and Telegraph Office, where demonstrations of how the coded message was sent were given (see Figure 5.2), along with the reading of news-paper headlines of the time. The Mayor and other locals dressed in period costume (Figure 8.1), and a local band played appropriate music. A number of descendants attended the celebrations, including Kathleen Scott's grand-daughter (Ryan, 2013). Other activities included the laying of a memorial plaque, a Scott-themed play, *The Night Visitors*, a Scott 100 Dinner and the *Scott's Last Expedition* exhibition at the Canterbury Museum. The commemo-ration was important for Oamaru's identity, linked to their exploration

Figure 8.1 Mayor of Oamaru in period dress at the Scott 100 Celebrations, February 2013
Source: J. Laing.

heritage, based on the speeches we heard, the enthusiasm of the crowds attending the different events and the number of shop windows that entered into the spirit with dioramas and memorabilia of the period, including Ken dolls dressed as Scott of the Antarctic (Figure 8.2) and replicas of the *Terra Nova*. It can also be summed up by the following comments by a local in the Letters to the Editor page of the *Oamaru Mail*: 'Oamaru has stood proudly on the world stage as we've recognised our small, yet significant, page of this heroic story' (Elliffe, 2013: A6).

There have been a spate of Shackleton re-enactments in recent times, perhaps reflecting that Shackleton is very much the *explorer's explorer*. Many of the people we interviewed were passionate about his influence and inspiration. Karen refers to him as 'my friend Ernest Shackleton, an absolute marvel', while Geoff identified with Shackleton's struggles to raise funds for his expeditions. Peter Hillary refers to him as 'the true rock star of the polar game ... part of my life education, handed down' (Hillary & Elder, 2003: 303). His leadership was his standout quality for many explorer travellers (see, for example, Arneson & Bancroft, 2003; Woodhead, 2003). He was also a hero to his own generation, who paid tribute to his courage and

Figure 8.2 Diorama in the shop windows of Oamaru to mark the Scott 100 Celebrations, February 2013
Source: J. Laing.

decision-making ability. When Amundsen stood at the South Pole, he mused in his subsequent account:

> We did not pass that spot without according our highest tribute of admiration to the man, who – together with his gallant companions – had planted his country's flag so infinitely nearer to the goal than any of his precursors. Sir Ernest Shackleton's name will always be written in the annals of Antarctic exploration in letters of fire. Pluck and grit can work wonders, and I know of no better example of this than what that man has accomplished. (Amundsen, 1912: 439)

In 2013, Australian explorer traveller Tim Jarvis led an expedition, the *Shackleton Epic*, to recreate Shackleton's legendary feats in 1914 (Figure 8.3). While sailing to Antarctica on a polar expedition, Shackleton's ship, the *Endurance*, was trapped in the ice. Once the ship sank, Shackleton and his men spent some time living on the ice, before it broke up and they managed to escape in lifeboats to Elephant Island. Shackleton then embarked on a

Figure 8.3 Tim Jarvis on the South Georgia crossing during the 2013 Shackleton Epic
Source: P. Larsen/Shackleton Epic.

rescue mission. He took five men with him on the lifeboat *James Caird*, sailing 1500 kilometres to South Georgia Island through perilous seas. He then made the first crossing of the island, which took 36 hours. Exhausted, but victorious, Shackleton sounded the alarm at the whaling station at Stromness, and then worked to evacuate the remainder of his team. Incredibly, no lives were lost. The key qualities associated with Shackleton and his miraculous journey to rescue his fellow expeditioners – bravery, persistence, mental strength – are manifested in many of the motivations behind explorer travel (see Chapter 2).

Jarvis was keen to maintain the authenticity of what was done back in 1914, even down to the gabardine coats they wore, the tools they used and a replica lifeboat: 'We have no intention of not doing things the way Shackleton did until we really have to . . . Shackleton relied on adaptability and for us it's the same' (Robertson & Darby, 2013: 1). Their only concession was to travel with a sealed box of modern emergency equipment (Robertson & Darby, 2013). The challenges included 140 kilometres per hour winds and snowstorms and saw three of the six re-enactors drop out (Tedmanson, 2013).

Figure 8.4 Tim Jarvis during the 2007 Mawson re-enactment in Antarctica
Source: M. McDonald.

This was not Jarvis' first re-enactment experience. In 2007 he recreated the expedition to the South Pole led by Sir Douglas Mawson (Figure 8.4), in which the latter's two teammates died, and 'used the same clothing, equipment and starvation rations as Mawson had available to him in 1912 to test the theory as to whether he needed to cannibalise his fallen colleague to survive' (Jarvis, 2013). His Shackleton re-enactment, in contrast, was less anchored in solving a mystery and more about paying respect to the valiant deeds of the past. As Jarvis noted as he reached the Stromness whaling station: 'It's left me with greater admiration for Shackleton and what he managed to achieve' (Tedmanson, 2013: B3).

Another Shackleton re-enactment in 2009, the Matrix Shackleton Centenary Expedition, led by Henry Worsley, vividly illustrates the hero worship that Shackleton continues to inspire.

In Shackleton's Footsteps: A Return to the Heart of the Antarctic (Henry Worsley, 2011)

Our first inkling that Henry Worsley is a Shackleton tragic is when he writes of the night he slept by the explorer's grave in Iceland. He longs to have been a part of 'the Edwardian heroic age of polar exploration' (Worsley, 2011: 2) and no hero evokes more admiration from him than Shackleton. He has become a mentor through his Army career and Shackleton's leadership style is one which Worsley tries to follow with his soldiers. He also has a genealogical link – Worsley is related to Frank Worsley, the captain of the

Endurance and one of the men on the *James Caird* who took part in the famous rescue, culminating in the desperate crossing of South Georgia Island. He notes that a fellow polar enthusiast once told him: 'Damn it, man – you must have the same genes' (Worsley, 2011: 6). He meets Shackleton's granddaughter Alexandra at an auction, who introduces him to Will Gow, her cousin, who is Shackleton's great-nephew by marriage. Will has plans to undertake a re-enactment of Shackleton's *Nimrod* journey, to mark the centenary, and Worsley is keen to join him. They are eventually joined by Henry Adams, who is the great-grandson of Jameson Adams, one of Shackleton's team of four who attempted to reach the South Pole in 1909, but turned back 97 miles from their goal.

Worsley is also intrigued by the prestige of what they are trying to do: 'More people had stood on the surface of the moon than had attempted our intended route' (Worsley, 2011: 42). He meets one of the latter, Robert Swan, and is given a teddy bear which he had carried on his North and South Pole journeys. His other treasured talisman is a compass which Shackleton used on his Nimrod expedition. Before he leaves, he reads everything he can on Shackleton, and his departure point on the ice is Shackleton's hut. As he later writes: 'I could hardly get closer to my mentor. The only thing left was to walk in his footsteps to the Pole' (Worsley, 2011: 97).

The team read extracts from Shackleton's diary each night and Worsley is constantly dwelling on his hero. He writes that he felt 'closely connected to the past' after one of these readings: 'Hopefully, tomorrow, we would see what [Shackleton] had described' (Worsley, 2011: 137). The modern expeditioners are annoyed by the intermittent intrusions of other human activity, such as the contrails of aeroplanes and people on snowmobiles or skis, but this quietens down as they get further into their expedition. Mostly it is just the three of them, in a largely all-white landscape. This makes it easier for them to imagine themselves on Shackleton's expedition and feel close to his team: 'We were looking at exactly the same piece of high ground that they had seen. It was not difficult to imagine them talking about which route to take to get them to the highest point' (Worsley, 2011: 129). Worsley sees two birds, the only living creatures besides them for miles around, and interprets this experience as spiritual and deeply significant: '[These] could be the spirits of Shackleton or Scott and this was a sign that they were keeping a watchful eye over us' (Worsley, 2011: 123). He later imagines Shackleton camped beside them, and would 'walk over to his tent and listen in on what they were discussing' (Worsley, 2011: 150). We are interested in Worsley's comment that he felt the presence of someone watching over them, and his reference to Shackleton's comment about a 'Fourth Man' (Worsley, 2011: 241) being present during his South Georgia traverse. Such psychological phenomena have been labelled *the Third Man Factor* (Geiger, 2009). Michael, one of our interviewees, also experienced it:

> The sensation of going along feeling as though I was almost in the company of someone else, which I often get when I'm on these trips, particularly if you're in a slightly mind-altered state. I know I was in the company of someone else in the South Pole but we'd often travel five, six, seven hundred metres apart at a speed you feel comfortable with to keep yourself warm and I'd find that I was really conscious of the fact that there was almost someone there with me. And I think it's the more resourceful part, having given it a lot of thought, the more resourceful part of my own personality that comes out in those situations. But it's one that my normal everyday self, whatever you want to call it, is not accustomed to being in the company of, something along those lines anyway.

The role play in the Shackleton re-enactment extends to enjoying the same treats at Christmas as Shackleton's team did – a cigar and a sip of crème de menthe. When the journey becomes tougher, these imaginings help Worsley to break through the mental barrier. He is amazed at how resilient the explorers were, and how their readiness to keep going was a consequence of Shackleton's positive and optimistic leadership style. At the point where Shackleton could go no further towards the Pole, Worsley and his team stage a photo where each stands in the spot of the original 1909 photograph: 'I was Shackleton, Will was Frank Wild and Henry was his great-grandfather, Jameson Adams' (Worsley, 2011: 196).

The journey is described as an 'utter privilege' (Worsley, 2011: 135) and a way to honour Shackleton and his team. As they get closer to the point where their forebears turned back, Worsley observes 'This journey was about them, not us. Now was not the time to feel the glow of pride for what we were about to complete. These final few miles belonged to them' (Worsley, 2011: 192). Their humility, however, had its limits. There is a touch of ego in the way Worsley refers to their crossing of the Great Ice Barrier as allowing them to join a 'list of extraordinary men' (Worsley, 2011: 132) and his need to carve his team's names on a small rock besides those of Shackleton's team: 'I wanted there to be a record of Will, Henry and I having passed this way as well' (Worsley, 2011: 136). They also get annoyed when their support team turns up to trek the last 97 miles that Shackleton missed, even though they were expecting this. Worsley is not proud of his reaction: 'We didn't want our space intruded; we wanted to savour our achievement by ourselves' (Worsley, 2011: 203). The authenticity of their *real* journey is contrasted with those who had merely flown in at the end. It resembles the reaction of Best and Bowles (2007) on the Camino Way, when groups join the pilgrimage close to Santiago de Compostela and some take the final stages by bus, instead of walking. This is a form of cheating in their eyes, and belittles their efforts and injuries.

The re-enactment ends with another spiritual moment for Worsley – a parhelion in the sky – in which 'some other force seemed at work' (Worsley, 2011: 227). He has taken Shackleton's compass to the South Pole, and fulfilled both men's quests. Worsley does, however, recognise the personal obsession that drove him to this undertaking and his 'total immersion' in the enterprise. Like Shackleton, his family were forced to 'stand and wait' for his return (Worsley, 2011: 242, 243). The recognition of the toll of these journeys on friends and loved ones is important, and something which will only increase as we head into the next space age – and potentially send human beings to Mars.

Part 3

Tourists at Play

9   Crossing Borders

In 1984, Nick Danziger received a Churchill Fellowship to follow the ancient Silk Road across Asia. However, rather than its history, he was interested in its modern dimensions, particularly the manmade barriers to travel. Danziger wanted to cross closed borders, taking on these challenges specifically because they were forbidden. His route would take him through Iran and Afghanistan and into Western China, all at that time closed to foreigners and highly dangerous.

In planning his solo Silk Road exploration, Danziger aimed for immersion in the cultures he would encounter:

> I would do the journey overland, using local transport, hitching and walking. I had no desire at all to join the present trend for travelling long distances in eccentric ways, like cycling or hiking all the way. My journey would be one of discovery, and its object would be to promote greater understanding of the people along the route; it would not be a journey about me; it would be a journey about them. (Danziger, 1988: 8)

Danziger saw himself as an outsider. With an American father and an English mother, he had grown up in Switzerland. Uncomfortable with his privileged surroundings and bored at school, as a teenager he embarked on a series of adventures across Europe. Without money or tickets, he journeyed to various cities, becoming adept at scamming tourists with hard luck stories. Later he worked as an interpreter and travelled through South America. For this trip, he constructed, one might even say fantasised, a persona of the wandering stranger: 'at times it would be to my advantage simply that I was a stranger, a foreigner, for people frequently find what is new and unusual more attractive and interesting than the familiar' (Danziger, 1988: 22).

A major challenge was finding the right balance between immersion and objective distance if he was to fully understand what he would experience.

This is the intriguing paradox we all face, whether we identify as explorers, travellers or tourists. As Danziger rationalised it:

> In any event, travelling as I would through countries in political turmoil, I was determined to keep an open mind, and to interpret what I saw rather than judge it. Certainly, if I was to succeed, I would have to stay enthusiastic, cool and persevering. I would have to use all the guile that had stood me in good stead in my former travels, and all the audacity too. (Danziger, 1988: 22)

Bizarrely, Danziger crossed a forbidden border into post-revolutionary Iran to find a long-lost tourism industry. A hotel desk clerk in Shiraz sought him out. Introducing himself as David, he explained that he was desperate to be able to speak English once again: 'like many Iranians who remembered the days of the foreign tourist boom he spoke with a mixture of nostalgia and pity about the good old/ bad old days of the sixties, remembering the cannabis trail and the long-haired hippies who travelled it' (Danziger, 1988: 82–83).

Crossing into Afghanistan, he joined the Mujahedeen (at this time the West was sympathetic to them in their war against the Soviets). Danziger adopted a disguise, pretending to be an Afghan beggar and masking his blond hair in a turban. While immersing himself in this role, Danziger was very conscious that he was only a visitor:

> I thought sadly of the day when I would have to leave my companions and continue on my way. For I always had the possibility of escape – of simply walking away when I had enough. These people were condemned to see their fate through. (Danziger, 1988: 142)

Once in Pakistan, he started to encounter Westerners again. He was very pleased with himself when some Australian tourists were rude to him; they had not seen through his disguise. In Islamabad, he donned another costume for deceit – ever the trickster, Danziger relished his theatrical escapades. In a suit and tie, he pretended to be a business executive and tricked the hapless staff at the Chinese Embassy into issuing a visa.

Danziger can be juxtaposed with Fred Burnaby, who also crossed borders in Central Asia. Burnaby was one of the great adventure travellers of the 19th century (and his wife Elizabeth Hawkins-Whitshed was one of the pioneers of female mountaineering and film-making). In 1875 Burnaby was a Captain in the British Army when he rode across the Russian Empire to Khiva in modern Uzbekistan. The impetus for this was that he had seen a newspaper article reporting that Russia had decreed that no foreigner would be allowed to travel in Russian Asia. Burnaby's self-appointed mission was to test this and demonstrate that an Englishman could go wherever he liked. In this sense, his journey was very similar in intent to that of Danziger just over a hundred years later.

For Burnaby, clothing was important, but here he differed from Danziger. While Burnaby was interested in observing and writing about exotic customs, he had no desire to immerse himself in their culture by wearing their clothes. Indeed, his quest required that he be clearly visible as a British officer on holiday. There was no point in defying the Russians if he did it in disguise. Burnaby was not an isolated case; half a century later, the explorer Henry Savage Landor proudly proclaimed that 'I did not masquerade about in fancy costumes such as are imagined to be worn by explorers' (quoted in Grann, 2009: 140).

Burnaby's travelling outfit was smart and practical, 'a suit of clothes, made by Messrs. Kino, of Regent Street, and in which they assured me it would be impossible to feel cold' (Burnaby, 1877: 13). Approaching Khiva, he sought out a barber for a shave – for the Victorian reader it was such cultural vignettes that were particularly savoured – and Burnaby milked it for the humour of the barber's attempts to convince him to have his head shaved as well. He also took off his sheepskin coat and donned a military black fur pelisse, 'so as to present a more respectable appearance on entering the city' (Burnaby, 1877: 294). To meet the Khan of Khiva, he wore his best, a black shooting jacket, and 'I had brought one white shirt, thinking that I might possibly have an interview with some Central Asian magnate or other. Greatly to my surprise, the article in question was not much the worse for the journey' (Burnaby, 1877: 304).

Like Burnaby, Danziger was very tall and clearly a Westerner. He relied on the tradition of providing hospitality to strangers. However, he also understood he could not be too strange. He wore clothes, particularly headgear, which helped him to partly blend in. His choices of apparel demonstrated empathy with the cultures he was encountering, plus they were practical in the harsh physical environments. Aiming to immerse himself in traditional cultures, to make friends, to dress and eat like them, Danziger followed a common pattern taken by many – that of crossing cultural borders to seek a profound experience.

Danziger's journey highlights an important aspect of being an explorer traveller. Rather than making geographical or physical discoveries, he was seeking out cultural encounters and trying to understand different societies and ways of life. This has always been part of the explorer's quest and is a strong motivation for modern explorer travel. In this chapter, we examine the intricacies and appeal of cultural immersion, the attempt by explorer travellers to blend in with the cultures they are visiting with an ultimate aim of greater insight. Our focus in this chapter is on those travellers whose aim is specifically to cross over cultural borders and immerse themselves in very different societies. For our purposes, we term them *immersive explorer travellers*.

Such immersion, we argue, proceeds along a continuum. For some, such immersion is fairly superficial, perhaps wearing local clothing and sampling regional foods. Others go deeper, even engaging in a fantasy that they can blend into a culture so well that they can pass off their disguise. Others, in turn, seek a potentially permanent transformation, exchanging their identity as a visitor for that of a new member of this culture. As with many aspects of the explorer traveller, there are elements of fantasy (even self-delusion) and a range of paradoxes and contradictions.

Getting the Costume Right

Among modern-day explorer travellers there is a seductive fantasy that they can fit in with exotic societies, particularly through the power of the right clothing and attention to detail. Putting on traditional costume is akin to donning a disguise, stripping off the trappings of their normal culture and substituting another in its place (Cronin, 2000). The process is transformative. It is a means of escape, perhaps even rebirth. Highly attractive for many, it is usually undertaken with the recognition that it is only a partial transformation and it is only temporary. This process is well illustrated through our interviews with two quite different explorer travellers.

Max, an Australian lawyer, was interviewed as he returned from Nepal. He was dressed in a Daura Suruwal (or Labeda Surawal), the traditional Nepalese outfit of a long tunic over pants. On his head was a Topi, a Nepalese brimless hat. He explained:

> I like to go to Kathmandu and dress like this. I don't have to wear my lawyer's suit. I don't have to shave. I can be myself ... It's important for me to wear it on the aeroplane coming back. I won't change until I have to go back to the office. I get hassled by Customs. They search me and give me a hard time. It's worth it. I don't have any worries about them ... I don't like being a lawyer, but it pays well. Going to Kathmandu and dressing like this is an escape. It allows me to keep going.

Wayne, when interviewed, was in New Mexico and was wearing a stereotypical cowboy outfit, including hat, boots, shirt and waistcoat. This for him was also a deliberate costume of choice:

> I'm a postman in Los Angeles. I deliver the mail and I'm just a postman. But, each year for my holidays I get dressed up like a cowboy and come to the West. I just hang around, drift from place to place. And I'm a cowboy ... I love Westerns. I've seen lots of movies. It's a look I love. Dressed like this, I'm a somebody. I'm a cowboy. Not a postman.

It is important to understand that both Max and Wayne saw themselves as wearing a convincing costume. These were not just souvenirs, but complete ensembles. Both were solo travellers. Both keenly did this as much as they could, planning regular holidays. Dressed to travel in their chosen destinations, they embraced different personas, which were constructed in opposition to their normal employment and lives. Their enjoyable travel fantasies allowed them to experience satisfying existential authenticity. Both professed that they did not care what others thought, although there were indications that they enjoyed negative reactions (Max, the lawyer, seemed to relish duelling with Customs officials). Both saw this dressing up as temporary, a break from the constraints of the everyday. Neither indicated that this was something they would like to do permanently. Their plans were for holidays, not a change of lifestyle.

Such practices may not be aimed at fooling the locals. Zurick, commenting on modern Western tourists wearing traditional Afghan clothing, argued that it was:

> More than a matter of taste among the Western travelers; it was a rite of passage. In discarding their Western clothes, travelers symbolically shrugged off their societies. In adopting native dress, travelers did not become natives; that would be impossible. Rather, they demonstrated to other Westerners that the passage from their own world to that of the 'other' had indeed been achieved, a passage that deserved recognition and certainly a new look. (Zurick, 1995: 66)

Dress-Ups for Grown-Ups

An obsession with costume suggests re-enactors and there is value in examining that phenomenon to understand the interest of explorer travellers in donning exotic clothing. Re-enactment is a form of living history in which participants dress and act as people from the past. It is a form of *serious leisure*, where participants develop their costumes, skills and accoutrements over many years and gain much of their satisfaction and self-identity from their success in doing so. It is a communal activity; re-enactors often come together in hundreds or even thousands and greatly enjoy the *communitas* of their interest (for a more detailed discussion, see Frost & Laing, 2013).

The importance that re-enactors place on the perceived authenticity of their clothing is well illustrated in a detailed study by Belk and Costa (1998). They looked at people who came together to recreate the lifestyles of Mountain Men – trappers and adventurers who travelled well beyond the Western frontier of the USA during the period 1825–1840. In a sense, these Mountain Men were explorers, although they rarely wrote down their finds, and a number – for example Kit Carson – were important guides for official exploring expeditions. At certain times of the year, the Mountain Men would come together to trade at events known as Rendezvous and it is these meetings that are re-enacted in the modern world.

Belk and Costa participated in 13 Rendezvous events. All were held in forested public land around the Rocky Mountains. Hundreds of participants travelled to these events and established short-term camping communities, dressing in period costume and engaging in traditional activities. Belk and Costa analysed these:

> Rendezvous participants socially construct and jointly fabricate a consumption enclave, where a fantasy time and place are created and experienced ... The modern mountain man rendezvous as a fantastic consumption enclave is found to involve several key elements: participants' use of objects and actions to generate feelings of community involving a semimythical past, a concern for 'authenticity' in recreating the past, and construction of a liminoid time and place in which carnivalesque adult play and rites of intensification and transformation can freely take place. (Belk & Costa, 1998: 219)

This transformation from the modern to the past takes serious time and effort and involves more than just putting on a costume. The Rendezvous are *social worlds*, with their own customs and traditions. Newcomers, known as *pilgrims*, are expected to participate in quite a number of events before being fully accepted into the group. Developing Mountain Men costumes and accoutrements is a slow and cumulative process. While gear can be purchased, part of the fantasy is engaging in ceremonial bartering as the Mountain Men did nearly 200 years ago. Great respect and status is gained by those who are adept at making their own authentically styled costumes. Others are valued for their expertise in archaic crafts such as flint-making, tanning hides, horsemanship and bead-work. A third vehicle for excellence is in exercising appropriate braggadocio during the carnivalesque evenings, outshining others in prodigious displays of swearing, vernacular language, tale-telling and drinking. Throughout the event, rituals and traditions, whether real or invented, allow the costumed participants to pass over into a liminal space where they are transformed (Belk & Costa, 1998).

Similar practices and customs are apparent at Helldorado Days, in Tombstone, Arizona (observed in 2012). This three-day event can be partici-pated in at a number of levels. At the deepest are costumed re-enactors (see Figure 9.1). As with the Mountain Men Rendezvous, they are engaging in serious leisure, creating their own social worlds, dressing as and acting out certain key characters. For men, the most popular is to *be* Wyatt Earp. Noticeably, the model is cinema's Kurt Russell (and not Kevin Costner). Great attention is paid to clothing, hairstyles (real moustaches, longish hair), hats, weapons and the right swaggering attitude. For women – and this is an unusual re-enactment event for the large numbers of females participating – the most common outfit is a dance-hall girl. Strangely the Western stereo-type of the School Marm is not at all popular, although there are a few buckskin-clad cowgirls. While there is interaction with onlookers, for

Figure 9.1 Helldorado Days re-enactment at Tombstone in 2012
Source: W. Frost.

example, posing for photographs, the various re-enactor groups form their own clans, tending to party and socialise among themselves. Again rituals and traditions are combined with costume to immerse the participants in a liminal space (Frost & Laing, forthcoming, 2014).

While re-enactors share some characteristics with the immersive tendencies of explorer travellers, there are some key differences. For re-enactors, adeptness at costume and character allows them acceptance into a social world based on a respect and interest in one's own cultural heritage. In contrast, the immersive explorer is usually quite solitary, seeking admittance not to a group of cultural peers, but to a completely different society. The immersive explorer is not a joiner, rather an outsider seeking a new and distinctively exotic group. The status they seek is internal – although there may be elements of showing off – through the individual shaping of a new self-identity. Not surprisingly, the typical narrative of the immersive explorer tends to begin with a strong sense of dissatisfaction with their current situation. Their travel then becomes an internal journey to find a better life. As Forsdick (2005) suggests, there may also be a sense that this exotic society and lifestyle is under threat and that this is a chance to experience this before it falls victim to modernity and globalisation.

A New Imperialism?

What is most contentious about the desire of some explorer travellers to completely immerse themselves in exotic cultures is that it is an unequal bargain. The common narrative is of a Western traveller – White, well off, educated, urban – who seeks to exchange their lifestyle for another. That other is typically constructed as better: simpler, traditional, untainted by the artificiality of modern life and more authentic (Forsdick, 2005; Polezzi, 2006). Except this is not a true exchange. What the Western traveller gives in return is a complex mix of good and bad. Even in terms of general tourism, less developed traditional societies face a range of inequities (Zurick, 1995):

(1)   The major tourism operators (airlines, hotel chains, travel agents, cruise operators, tour companies) are large-scale commercial entities based in Western economies. The revenue that flows to them primarily stays in the First World.
(2)   Employment for traditional peoples may be limited: either low-paying manual labour or in many cases non-existent. Management and skilled jobs tend to be occupied by outsiders. For example, the resort of Yulara was established in 1976 to service visitors to Uluru in Central Australia.

However, when in 2011 it was sold to the Indigenous Land Corporation, only two of its 700 staff were from the local indigenous community.

(3) Tourism destinations and operations in the developing world are *price-takers*. They compete against a wide range of offerings which may be easily substitutable. This allows Western operators to keep fees and costs low.

(4) While national parks are controlled by individual countries, they are usually highly centralised. Accordingly, local people may have little say in development decisions and revenue may flow to urban capitals.

(5) Souvenirs may be copies of traditional handicrafts, made cheaply overseas and pricing out local artisans.

(6) The relative wealth and range of possessions of foreign tourists may draw local peoples into global markets. A common effect is to encourage the drift towards tourist cities – the *honey pot* effect – where they become trapped in low-paid menial jobs.

In addition to these issues, explorer travellers may also be responsible for further impacts. In particular, their economic benefit may be low and infrequent. The immersive explorer traveller is typically trying to lessen their expenditure: travelling light and seeking out cheap accommodation and meals. Danziger, for example, was very proud that he hardly spent any of his Churchill Fellowship. Even at this low level of expenditure, there is still a commodification of customs and traditions, with travellers demanding performances and insights in return for gifts that are little more than old-fashioned trinkets. In taking advantage of traditional hospitality, they may exhaust scarce resources while providing little economic benefit. In short, the explorer traveller immersing themselves in a traditional society may be just practising a new version of cultural imperialism (Cronin, 2000).

To examine this unequal exchange and other issues arising from immersive explorer travel, we conclude this chapter with a selection of examples of explorer traveller narratives into Australia, Africa and Asia.

Australia

The ballad of John Wilson

The 19th century explorers had problematic relations with the local Aborigines. Burke and Wills are famously remembered for warning off curious Aborigines by firing guns in the air. Suffering greatly from exhaustion, a lack of food and vitamin deficiencies, it has been argued that forming closer relationships would have saved them. In contrast, Paul Strzelecki, lost in dense rainforests that now bear his name, only survived through the ability of his native guide Charley Tarra to catch koalas for food.

The most extraordinary example of an explorer immersing himself in indigenous culture was that of John Wilson. Transported as a convict on the First Fleet, he was released in 1792. Crossing over the frontier, he *went native*, living with local tribes and adopting the name Bunboee. In 1796 he applied for permission to provide for himself (that is to subsist outside of the colonial economy). This was refused and Wilson was threatened with the prospect of being declared an outlaw. Fearing that their example might encourage convicts to abscond, Governor Hunter ordered Wilson and three others to return to civilization. Wilson walked into Parramatta wearing only a kangaroo-skin apron and with the marks of ritual scarification on his shoulders and chest.

Reasoning that a poacher makes the best gamekeeper, Hunter recruited Wilson for an unusual expedition. Among the convicts there were rumours of Whites living a free and easy life in good country beyond the seemingly impassable Blue Mountains. To scotch such talk, Hunter selected four leading convicts to make the journey and see that they were wrong. Wilson was appointed to guide them. Three of the convicts quickly gave up, but the remaining one persevered with Wilson. The small party promptly crossed the Blue Mountains and found promising country. On a second expedition Wilson repeated his success (Cunningham, 1996).

Australia's national narrative mythologises Blaxland, Lawson and Wentworth as first crossing the Blue Mountains in 1813. Wilson's successful expedition in 1799 is hardly known about (and certainly not taught in schools). Cunningham (1996) argues that Wilson failed to get due credit as his journey achieved the opposite of what was intended. The authorities wanted proof that the mountains were impenetrable. When Wilson succeeded, they kept it quiet. While agreeing with Cunningham, we also see another reason why Wilson has been painted out of the historical canvas. The explorer myth requires stereotypical heroes – officers and gentlemen like Blaxland and his companions. A White ex-convict turned Aborigine does not fit these popular conventions.

Most importantly, Wilson's relationships with the Aborigines gave them no long-term advantage. Indeed, it may have accelerated conflict and the spread of disease. Neither Wilson nor any of the other ex-convicts became useful mediators between the two societies.

Cinematic encounters

More recently, cinema has been a great influence on how travellers see the Australian Outback and encounters with the indigenous people. Taken as a group, films about the Outback present it as a place of potential transformation, where the visitor will be tested and changed. Part of this is due to the wilderness landscape and part to the interaction with Aborigines (Frost, 2010). Such cinematic representation mirrors Western ideas that there is a special spiritual or magical quality to the Outback which can profoundly affect certain visitors, especially if they are open to new experiences. Accordingly, travelling to the Outback is often seen in terms of a spiritual pilgrimage (Digance, 2003).

Positive interactions with indigenous peoples are a much stronger motif in Australian cinema than Hollywood Westerns. *Jedda* (1956) had Aboriginal leads at a time when the USA still used White actors to play Native Americans (as in the German actor Henry Brandon playing the Comanche warrior Scar in *The Searchers*, released at the same time). Hugely successful, *Jedda* promoted a radical message that Aboriginal culture was strong and powerful and that forced assimilation was unworkable. However, it has elements of a Shakespearean tragedy; the young Aboriginal lovers are fated to die, as they cannot fit into European society.

Other films have continued with this positive representation, often mixing it with the idea that special visitors can somehow gain access to Aboriginal culture. In *Walkabout* (1971), an English teenager (Jenny Agutter) is lost in the desert. She is rescued by an Aboriginal boy (David Gulpilil) who is travelling alone as part of his initiation ceremony. He helps her survive and guides her to safety. Similarly, in *Eliza Fraser* (1976), an Englishwoman (Susannah York) is shipwrecked in the early 19th century and survives by being adopted by Aborigines. In *Crocodile Dundee* (1986), an American photojournalist (Linda Kozlowski) comes to Kakadu to file a story on tour guide Dundee (Paul Hogan). While the eponymous hero is very much constructed as an Australian Davy Crockett, it is also revealed that he is a *White Aborigine* – fully initiated into the local tribe and engaging in their secret ceremonies. In *Australia* (2008), Lady Sarah Ashley (Nicole Kidman) travels from England to the Northern Territory. Initially cold and uptight, she is gradually softened by her relationship with a young Aboriginal boy. As fiction, these films have *feel-good* elements, promoting a view that immersion is possible and desirable, but any negativities are glossed over.

Mutant Message Down Under (Marlo Morgan, 1991)

While parts of these storylines make some Australians squirm, the most controversial imagining of cultural immersion in the Outback was the novel *Mutant Message Down Under* by Marlo Morgan. This purported to tell a true story and was even designated by its publisher as *non-fiction* until public pressure forced them to drop that label. Even now, it still contains an introduction that states that it is based on an actual experience and is only sold as a novel to protect the Aboriginals involved from legal action. Widely criticised by Aboriginal groups, there is no evidence of its veracity. Nevertheless, it was highly popular in the USA – claims of sales of a million copies have been made – and still seems to be widely accepted there as factual (Hiatt, 1997).

Morgan is an American health worker who visits Australia to set up training programmes. When she receives an invitation to meet with an Aboriginal group, she strangely assumes that it will be some sort of awards lunch. Instead, she is taken into the desert, where her clothes and belongings are burnt. Now she is forced to wear simple rags. She is instructed that she will accompany the tribe on a *walkabout* for three months. One of the Aborigines explains to her that:

> My people heard your cry for help . . . You have been tested and accepted. The extreme honor I cannot explain. You must live the experience. It is the most important thing you will do in this lifetime. It is what you were born to do. (Morgan, 1991: 15)

She gradually learns that she is with a tribe of full-bloods who still live a traditional life and are adept at hiding from the modern world. They call themselves the *Real People*, whereas Whites are *Mutants*. Travelling with them across the hostile environment, she realises the futility of the Western obsession with possessions and its selfish lifestyle.

In the desert, food should be a problem. However, as with many aspects of existence, Marlo grasps a different truth:

> I was told that plants and trees sing to us humans silently, and all they ask in return is for us to sing to them . . . The primary purpose of the animal is not to feed humans, but it agrees to that when necessary . . . each morning the tribe sends out a thought or message to the animals and plants in front of us. They say, 'We are walking your way. We are coming to honor your purpose for existence.' It is up to the plants and animals to make their own arrangements about who will be chosen. The Real People tribe never go without food. Always the universe responds to their mind-talk . . . When a snake appeared on our path, it was obviously there to provide dinner. (Morgan, 1991: 51–52)

Marlo participates in a *dream-catching* ceremony, where she enters a trance. She learns that 'the tribal people do not dream at night unless they call in a dream . . . [whereas] Mutants dream at night because . . . we are not allowed to dream during the day' (Morgan, 1991: 115). The next morning she realises she has been given a message to take back home:

> I learned that a time had come where I could no longer stay with the people, the location, the values and beliefs I held. For my own soul growth I had gently closed a door and entered a new place, and entered a new place, a new life that was equal to a step up a spiritual rung on a ladder ... If I simply lived the principles that appeared to be truth for me, I would touch the lives of those I was destined to touch. (Morgan, 1991: 116–117)

Mutant Message is preposterous and shallow. Its cultural insensitivity ranges from an obsession with full-bloods versus half-castes, to the transfer of Native American beliefs and customs to a completely different ethnic group. Nonetheless, it is a telling illustration of the fundamental problems of the immersion narrative. At its core it is a story of cultural appropriation – the claim that the heroine has been accepted into and *understood* another culture. In the case of this novel, it can be laughed off, but in a wide range of travel narratives and fiction dealing with immersion, this claim of ownership sits very uneasily.

Africa

Here we consider two well-known factual accounts of *settlement* by Europeans in Africa. Both are very popular and influential, and they provide quite contrasting narratives.

Out of Africa (Karen Blixen, 1937)

In 1913 the Danish Blixen moves to Kenya to establish a coffee plantation, remaining there until 1931. *Out of Africa* is a loose collection of stories about running the plantation, interacting with local tribes and her relationship with an English safari tour operator, Denys Finch Hatton. Blixen's goal is settlement, the establishment of a commercially successful farm. She is not attempting to cross over and live like the local people. Instead, much of the time she is reflecting on the cultural divide and the impossibility of immersion, as she holds a position of almost feudal power and dominance. She is intrigued that, while the locals rely on her for employment and assistance and she is constantly in their gaze, 'I reconciled myself to the fact that while I should never quite know or understand them, they knew me through and through' (Blixen, 1937: 26). She rationalises this distance, not in terms of colonialism (which, of course, she sees as natural and desirable), but as due to different stages of development, reflecting that, 'perhaps the white men of the past ... would have been in better understanding and sympathy with the coloured races than we, of our industrial age, shall ever be' (Blixen, 1937: 153).

The White Masai (Corinne Hofmann, 1998)

It is love at first sight when Corinne meets Lketinga. She is a Swiss bridal-wear shop owner on holiday in Kenya. He is a Masai tribesman who has just arrived in the city of Mombassa, seeking work. Marco, her boyfriend, points him out:

> It's as if I've been struck by lightning. A tall, dark brown, beautiful man ... My God, he's beautiful, more beautiful than anyone I've ever seen. He is wearing almost no clothes – just a short red loincloth – but lots of jewellery ... His long red hair has been braided into thin braids, and his face is painted with symbols ... the way he holds himself, the proud look and wiry muscular build ... I can't take my eyes off him ... he looks like a young god. (Hofmann, 1998: 2)

Corinne's boyfriend jokes at first that Lketinga has bewitched her. Completely obsessed, she abandons her life in Europe. She quickly finds difficulties with Masai culture and traditions. There are taboos against kissing and men and women eating together. Corinne laments that, 'all my romantic fantasies of cooking and eating together out in the bush or a simple hut collapse' (Hofmann, 1998: 24). After they are married, Lketinga is constantly jealous, especially when other men talk to Corinne.

The Masai are in a state of transition. The Kenyan government wants to end their nomadic life. Corinne is symbolic of the changes that are occurring. She brings vast wealth into the village, buying a car and setting up the first Masai-operated shop. It is *her* narrative, not *his*. This commercial activity makes sense to her, but we also get glimpses of the impact on Lketinga. As a warrior/tribesman, how does he fit in with this new world? It is Corinne who earns the money, not him. Her car and shop bring status, but he clearly feels marginalised. At the beginning, he has never drunk alcohol, but quickly becomes a prolific beer drinker. When he is in dispute with another Masai, he is mortified that she intervenes. Corinne constantly argues with him for wandering off on men's business and not doing what she tells him to do.

This is a very curious book. There is never a sense of getting deeply into Corinne's psyche. She seems to make rapid life-changing decisions without much thought or reflection. There is little insight into what drives her. There is no back story that might explain her actions. She does not seem to understand that there will be cultural differences between herself and Lketinga. She constantly rails against Kenyan bureaucracy, yet much of it – visas, work permits, car registration, concerns about bogus marriages – is quite common. She would certainly have similar bureaucratic issues in Switzerland. Of all the travel narratives we read for this book, this is the one we both felt was the most difficult to engage with. It may be that something is lost in the translation from German to English, although as we both speak German, we do not get a sense of that. Or it may be that Corinne keeps part of herself back from the readers. Certainly she comes across as narcissistic – and this may be a big part of the appeal for readers of what is a very popular book.

Asia

Asia, particularly South Asia, has wide appeal as a place where intrepid travellers may have spiritual and transformative encounters. Howard (2012: 138) argues that films, novels and travel stories *promise* life-changing experiences for those travelling to the Himalayas, 'a frontier zone *par excellence*'. Nonetheless, South Asia may be a confronting place, as is illustrated in the travel narrative of Robyn Davidson.

Desert Places (Robyn Davidson, 1996)

Davidson made her name with *Tracks* (1980), her account of a journey by camel across Australia (which has now been filmed, to be released in 2014). Highly popular, Davidson has a cult following as an author, partly due to three elements. These are her gender, her strong narrative of personal transformation through travel, and her intense self-reflection – sometimes coming across as self-doubt. Combined, these make her writings very personal and emotional and this clearly strikes a chord with many readers.

In *Desert Places*, she hatches the idea of accompanying the nomadic Raika people on their travels across the Thar Desert in Rajastan, India. She is very conscious that, while a solitary and nomadic lifestyle appeals to her, it is also fast disappearing in the modern world. Accordingly, her exploration has two purposes – to satisfy her own yearnings for immersion in an exotic society and to document a culture before it vanishes. The latter suggests the trope of 'last chance to see' tourism and the elegiac qualities of much travel writing (Forsdick, 2005).

The right appearance provides difficulties for the fair and blonde Davidson:

> For this project I had chosen neck-to-ankle-to-wrist cream cotton kurta pajamas. My hair was pulled back in a bun and I wore neither make up nor jewellery. I looked as alluring as a cow pat. (Davidson, 1996: 21)

Even then, her costume excites comments about the decline of traditional standards:

> I had thought my clothes the latest word in modesty but she [one of the Raika] suggested I cover myself more, that I do up the top button on my kurta, that I wear an orni, that I hide the string of my pajamas from view. She said, 'Everything has changed in India. People have no fear any more. And they are fed on those dreadful Hindi films. The Raika go to Pushkar and see western girls, drug addicts, going off with taxi wallahs for fifty rupees ... You must be *careful*!' (Davidson, 1996: 26)

A professional writer and traveller, Davidson is scornful of tourists, although she recognises at times that there is a little bit of the everyday tourist in her. She laughingly reflects on how she shares some of the fanciful notions that privileged Western tourists carry:

> Somewhere in the small residue of my romanticism I had been harbouring the delusion that dacoits were Robin Hood types with attractive moustaches and incipient socialist tendencies, who might, if encountered, be interesting to chat to and who would at the very least provide good book fodder. (Davidson, 1996: 68)

She is also not immune from the need to collect souvenirs. In a village shop just before the Indian festival of Divali, she buys 'some firecrackers for myself, as the packaging sported a portrait of Chairman Mao surrounded by exploding crackers and Indian gods. It was a prize of kitsch' (Davidson, 1996: 154–155).

Her journey is physically and emotionally tough. This is not a wilderness like Outback Australia; everywhere she encounters crowds and abject poverty. Her clothes take on a different meaning. They do not really help her to fit in, but they can partly at least, offer escape. In a new village, 'to avoid an inevitable collection of stares, at least for a few minutes, I pulled an orni [veil] over my face [and] sank low into the seat' (Davidson, 1996: 46).

The relationship with the Raika is complex, sometimes close, often strained and not what Davidson had expected. Not as traditional as she anticipated, she sees that they are happy to benefit from their association with her. Taking a child to a hospital, they are 20th in the queue. Then the doctor sees there is a Westerner in the waiting room and they are called in straight away: 'For ever after ... [they] would want me to accompany them to clinics' (Davidson, 1996: 114).

Davidson had been proud that she was non-judgemental of other cultures. However, the poverty, bribery and crowding wear her down:

> But lately, battalions of judgements had been arriving unbidden from some less evolved self and nothing I could do would fend them off ... this new species of thought frightened me. What would I become if I allowed them their space? Would I become the kind of person I despised? (Davidson, 1996: 124)

> What did other people see that I could not see? Or what was I seeing that others apparently did not see? The well-off westerners who came here for enlightenment, what did they see? This repression and corruption, [did they see that as] spiritual? The worship of the rupee, non-materialistic? This stasis, this dead end, a goal to be reached by humanity? (Davidson, 1996: 174)

Like few other travel writers, she bares her dark feelings. What should be a great adventure has degenerated into a horrific test of her self-identity. She writes of a rage that 'bubbled up from somewhere deep and was too strong to staunch. The words "I hate India" did not fit with the person I thought I was' (Davidson, 1996: 175). Completing her journey is anti-climactic. There is no sense of achievement.

Venturing into Unsettled Territory

More than any other group of travel narratives we consider in this book, we find those in this chapter the most unsettling. It is not that they tell of Westerners so dissatisfied with their lot that they travel searching for an alternative. Rather it is that once set upon their quest, they quite happily appropriate the cultures of others. For most there is seemingly very little self-reflection. They want it; they take it. That is their right. For all their rhetoric of understanding and searching for a better and more authentic culture, they come across as neo-colonials.

Burnaby and Blixen, of course, are part of an older imperial world. Burnaby journeys to Khiva as a British Officer to demonstrate that the

Russians cannot tell an Englishmen what to do. In the jockeying between empires, the locals count for little. Half a century later, Blixen settles in Kenya as a colonist, seeking to build a rich estate and regain status and wealth. Neither are interested in immersion; they observe the curiosities of native culture, but do not want to join in. Class dominates these colonial narratives. The shadowy Wilson *goes native*, but he is just a former convict.

In the modern world, the former colonies are now all independent. However, Westerners still travel with a strong sense of superiority. Nobody asks permission. It is their right to travel and prise open the doors to other cultures. Danziger wants to cross forbidden borders, justifying this with notions of increasing understanding and armed with his Churchill Fellowship. Getting through at all costs means deceit and trickery – but that is not really bad (in his eyes) – as this is the only way to achieve his goal. Morgan insensitively mashes together Aboriginal and Native American culture, her fiction masquerading as a true story. Hofmann is single-minded in her entry into Masai society; the most narcissistic of these travellers, she seems either oblivious to or uncaring of her impact. Davidson, in contrast, is the most reflective of all of them and has her values deeply challenged by her odyssey. Nonetheless, it was her decision to join with the nomadic Raika – again for understanding – but also for a story with *National Geographic*.

Whatever the effects upon the explorer travellers, their immersion has clear and unsettling impacts upon the cultures they venture into. There is a romantic fantasy that these are timeless cultures, unchanged, settled and stable. This is nonsense. All are undergoing massive social changes and it is their interaction with the modern world that allows the travellers to intrude. Wilson, Burnaby and Blixen enter into newly conquered territories. The imperial powers are expanding; local networks and relationships are crumbling. Danziger's borders are only recently closed. In Iran, for example, he meets a local who worked in a thriving tourism industry only 10 to 20 years earlier and, 20 years later, tourism to Iran has revived. Lketinga, the object of Hofmann's affections, has only just moved to the city, where he is trying to make a living dancing at tourist resorts. His is a universal tale of abandoning the rural lifestyle for the promises of the big city. Davidson similarly enters a traditional nomadic society that is just about to disappear.

These narratives combine cultural appropriation, the right to venture wherever they want, the urge to document the last days of cultures, the blind romanticism and downright self-centeredness. It is a disturbing mix. Immersing oneself in exotic cultures is a seductive part of the explorer myth, but it also highlights perhaps the ugliest aspects of the explorer traveller.

10 On Safari

The word Safari comes from the Arabic *Safarīyah*. It means *a journey*. Most Arabic words that have entered into the English language did so through Mediterranean trade and diplomacy; in contrast, safari originates through another path. It came out of East Africa in the 19th century, where it had passed into Swahili through Arab expansion and trade southwards. With the European race to carve up Africa in the late 19th century, it became part of the colonial lingo. As a result of its late adoption, the word safari is the same in all the languages of the European countries which had colonies in Africa: England, France, Germany, Italy and Portugal.

According to the Oxford English Dictionary (www.oed.com), the term safari was first used in 1859 by explorer Richard Burton in a journal account of his travels. This was followed in 1871 by David Livingstone using it in a journal account. It is noticeably absent from Jules Verne's *Five Weeks in a Balloon* (see Chapter 6). Despite being used by early explorers, the word was still unfamiliar to many, so Theodore Roosevelt, in writing an account of his hunting trip to Africa in 1909, felt he needed to carefully explain its meaning. It was probably Hollywood that fixed the term and the concept into the consciousness of the general public (Cameron, 1990).

Safari was an East African word, which was gradually applied throughout the continent. In South Africa, the term *Trek* held a similar meaning. From Afrikaans, it also described a foot journey, although perhaps with the difference of involving oxen rather than porters. Crossing over into English at roughly the same time as safari, it resulted in a 19th century linguistic divide between Southern and East Africa. In Henry Rider Haggard's *King Solomon's Mines* (1885), for example, trek is used throughout and safari is not.

Initially, safari was used to describe trading or exploring journeys. Late in the 19th century it developed more of a leisure focus. The elite of the European colonists had a passion for hunting – a simple transference of a high-caste leisure activity from Europe to Africa. Safaris would last for days, even weeks, as groups of colonists and European visitors searched for abundant game. By the early 19th century, commercial firms such as Newland and Tarlton were organising safaris for tourists. Accompanying the White hunters were large numbers of Africans, working as guides, beaters, porters and servants. Here was a potent symbol of European dominance of both the natural environment and indigenous societies (Beinart & Hughes, 2007; Cameron, 1990). The safari was more than just a hunting expedition. It developed its own rituals and style, of formal dressing for dinner and sipping champagne in camp.

Probably the most famous hunting safari was that of former US President Theodore Roosevelt in 1909. The larger than life Roosevelt is lionised in the USA for his cowboy persona, commitment to conservation and championing of the Museum of Natural History – the epitome of the dynamic explorer traveller (see Figure 10.1). Roosevelt had been guided through Yosemite by none other than John Muir, resulting in his decision to shift Yosemite National Park from state to federal control. On his African safari, Roosevelt was accompanied by the cream of big game hunters, including Frederick Selous – the inspiration for Henry Rider Haggard's Allan Quatermain. However, this safari was criticised for the excessive slaughter of wildlife, with 11,000 animals killed, including over 500 big game animals (elephant, hippopotamus, rhino) (Cameron, 1990). As with many of these early hunting safaris, the slaughter of wildlife was justified as somehow scientific and accordingly worthy (Beinart & Hughes, 2007).

Figure 10.1 Theodore Roosevelt statue, American Museum of Natural History, New York
Source: W. Frost.

In the 20th century, a further shift in the meaning of safari occurred, mirroring changes in African tourism. While hunting was valued as an iconic pursuit of the upper classes, it was threatened by overhunting coupled with poaching and the expanding indigenous population. Colonial officials sought out ways to conserve animal numbers. Game reserves were established in the 1880s and 1890s, leading to national parks in the 1920s (and in these early periods the terms were often interchangeable). Early national parks, particularly Kruger National Park in South Africa, had twin roles of conservation and tourism. Indeed, they were popular from the start. Kruger, established in 1926, had 800 visitors in 1928, but by 1938 received 38,000 visitors. These new tourists wanted to see animals close up, not hunt them. Furthermore, whereas game wardens wanted to exterminate predators, these were exactly the ferocious animals that the tourists wanted to see and tick off their lists (Carruthers, 2009; Frost & Hall, 2009). As more national parks were established after WWII, the concept of the safari evolved. It was still a ritualised journey for White leisure, but now the object was seeing and photographing the fauna.

In the modern world, safari is used to describe a range of organised tours. In Africa, the focus is strongly on wildlife viewing (Figure 10.2). Although the purpose of the safari has changed, what is presented to tourists is very much steeped in the traditions and rituals of exploring (Cameron, 1990). While the path is well worn, there is an aura of discovery and adventure. There is danger from lions, elephants and hippopotami. There is still the paradox of combining an adventure trip with the trappings of luxury, as epitomised by fine dining in the bush or champagne on a balloon safari. A major tourism sector,

Figure 10.2 Safari tour in Kenya
Source: K. Williams.

packaged up for high-spending international tourists, the modern safari epitomises how the explorer myth pervades modern tourism.

The term trek has also undergone a modern transformation. Whereas safari is associated globally with wildlife viewing tours, trek has become linked with organised walking tours into remote places. Nepal and the Himalayas have become particularly associated with Western tourists undertaking commercial treks. As in Africa, this has a major impact on the local economy and is exposing more traditional societies to global influences. The Himalayas are also the setting for many commercial mountaineering expeditions, although as with treks and safaris, such tourism products are now spread around the world.

Safaris, treks and mountaineering expeditions are paradoxical tourism experiences. They are deeply rooted in the myths and rituals of the intrepid explorer, offering the opportunity to emulate the heroic adventurer. However, they are also organised tourism products, commodified and commercialised. Explorer travellers, as we have seen, are reacting against mass tourism, yet the concept of a guided exploring tour seems to smack of the basic elements of what they are rejecting. Our aim in this chapter is to examine this paradox, considering the appeal and features of organised expeditions and particularly focusing on the uncertain positions of local peoples in servicing these constructed fantasies.

The Group Tour

Given their prominence, group tours have been given surprisingly little attention in research into the tourism experience. This is probably as they are viewed as a down-market and rigidly formulaic product, provoking strong feelings of cultural elitism among academic researchers (for a discussion of this bias against mass tourism, see MacCannell, 1976; Risse, 1998). There are, however, exceptions. The most prominent of these are small-scale ecotourism tours, which have been the subject of a great deal of research and are usually distinguished from the conventional mass group tours.

Group tours have the following common characteristics (Cohen, 2004; Frost, 2004):

(1) They last for more than one day and include overnight accommodation, meals and related services.
(2) They are pre-organised and follow a set itinerary. A paid group leader or guide accompanies and leads the group. There may be ancillary staff, including language interpreters, specialist guides, drivers and porters.

(3)   The tour is packaged and marketed around clear geographic and/or activity-based themes. Examples include an Amazonian adventure, a Nepalese trek, a wildlife-viewing safari and an Arctic cruise (Figure 10.3).

(4)   Places in the group are purchased by individuals, couples or small groups. Accordingly, the tour group includes a range of people who have not previously had contact with each other. For the duration of the tour they form a temporary society, interacting with, even relying on, each other.

(5)   Cohen describes a participant as 'the least adventurous kind of tourist'. A passive consumer following a well-prepared itinerary and protocols, 'he makes almost no decisions for himself and stays almost exclusively in the micro-environment of his home country'. On Cohen's (2004: 39) continuum of tourist roles, 'familiarity is at a maximum, novelty at a minimum'.

Individuals choose to join group tours, as opposed to planning and staging their own independent journeys. Some of the key drivers of choosing group tours are as follows (adapted and extended from Frost, 2004; see also Arnould & Price, 1993; Poudel *et al.*, 2013):

(1)   *Access.* The tour operator may have exclusive access to a place, either being the owner, custodian or having negotiated monopoly rights. Or it may simply be that a tour offers the easiest means for accessing a remote place.

Figure 10.3 Zodiacs off Spitsbergen near the Arctic Circle
Photo: K. Williams.

(2) *Expertise*. This includes knowledge of the local area, particularly its language, customs, vagaries and risks. For tours with an adventure component, for example, kayaking or abseiling, the operator provides equipment and a high level of knowledge in training, supervising and managing the risks for inexperienced tourists.
(3) *Interpretation/mediation*. Another area of expertise is interpretation – the provision of information that allows the tourists to understand what they are experiencing. This may be conceptualised as a mediation between the tourist and the unfamiliar host community and environment. Group tours provide a high level of interpretation, with guides generally seen as more personal, interactive and effective than signage and other fixed interpretation.
(4) *Convenience*. As the tour operator organises everything for a fixed price, the tourist is freed from such worries and can relax. This is particularly attractive to tourists who are money rich but time poor, and to those who are inexperienced travellers.
(5) *Company*. It may be more pleasant to travel and socialise with a group.
(6) *Engagement with traditional communities*. In some circumstances, tourists may choose to travel with a tour organised by local indigenous or traditional communities. In doing this they are seeking to engage with those communities and provide a direct financial benefit to them (even if this costs significantly more than alternative means of visiting the area). Such a scenario is intriguing in that it goes against much of the received wisdom about group tours and we will consider this in detail later in the chapter.

As a subset of group tours, safaris and treks share many of these features, raising paradoxical issues of the level of control versus expectations of adventure and individual discovery based on the explorer myth. These dichotomies are further magnified when we also consider the recent trends towards the managed staging of tourist experiences.

The Experience Economy

The concept of the *Experience Economy* was developed by Pine and Gilmore (1999) and sits within a broader research discussion of tourist experiences (see, for example, Arnould & Price, 1993; Frost & Laing, 2011; O'Dell, 2005; Ooi, 2005; Ritchie & Hudson, 2009). The central premise of the Experience Economy is that, whereas in the past businesses provided goods or services, the profit-making potential in the future is in providing experiences. Pine and Gilmore define these experiences as 'a series of memorable events that a company stages – as in a theatrical play – to engage [consumers] in a personal way' (Pine & Gilmore, 1999: 2). This metaphor of performance is used consistently throughout the concept. The place of business, whether a restaurant, attraction or public area, is a *stage*; products are *props*, staff are *actors* and the experience is *scripted*. Even surprises can be scripted, with Pine and Gilmore advocating this as 'staging the unexpected' (Pine & Gilmore, 1999: 96). It is noteworthy that this metaphor of guiding operators staging a performance predates Pine and Gilmore. For example, Arnould and Price (1993: 24) wrote that 'the guide is an impresario who facilitates the enactment of vaguely familiar cultural scripts, helping participants to transform experiences into treasured, culturally construed memories of personal growth'. However, the key difference is that while Arnould and Price saw this as more organic, Pine and Gilmore emphasise that it is carefully planned.

The Experience Economy was not specifically developed for tourism. However, it has become popular within various sectors and is now widely used and understood. Indeed, at times we have been fascinated by industry practitioners being able to fully describe the concept and explain how they have adopted it within their business model, yet not be able to put either a name to it or identify its two originators.

The concept is particularly applicable to commercial safaris, treks and other group tours which utilise the explorer myths. As noted above, 19th century safaris were highly theatrical and playful. Participants easily slipped into playing attractive roles such as the Big Game Hunter. A safari was staged with specific rituals, costumes, props and protocols. Much of this survives today, making safaris an attractive experience very much in the mould envisaged by Pine and Gilmore.

However, the experience economy has been subject to criticisms. It is important to consider these, as in the context of safaris and treks, they highlight the paradoxes of packaging the experience (or fantasy) of being an explorer within the structure of a group tour. The five main criticisms are as follows:

(1) Pine and Gilmore argue that the visitor experience has to be tightly scripted, in order to maintain a consistent quality throughout repeated deliveries. Without that adherence to a script, they argue, the operator would lose control over quality and there would be variations in customer satisfaction. However, an experience stripped of spontaneity and uncertainty may be unsatisfying. Hom Cary observes that memorable tourist experiences or 'moments' are often serendipitous and arise when

there is a 'spontaneous instance of self-discovery' (Hom Cary, 2004: 67). Scripting a performance in a remote and wild place seems incongruous, as 'the experience of the sublime and of undisclosed mystery is less easily packaged and sold as a product' (Lengkeek, 2002: 192). Furthermore, close scripting leads to rapidly diminishing levels of satisfaction for repeat customers. A trite example, but worth considering, is when one of us (Warwick) took his children on the Jungle Safari Ride at Disneyland. Bathing in a nostalgic glow, he loved every minute of it, particularly the corny banter of the guide. Deciding to do it again straight away, he realised with a thud that every part of the commentary, even including bad puns, was completely scripted and was unchanged at each performance.

(2) Pine and Gilmore stress consistency of experience, controlled through tight design, scripting and training. Ooi (2005) contends that such packaging of experiences is problematic. Experiences, he argues, are multifaceted – as are tourists – and individual tourists will have widely different responses to the same activities. This raises the marketing concept that modern experiences are *co-constructed* or *co-created*, with the consumer's responses and behaviours making a significant contribution to what unfolds (Frost & Laing, 2011). It is notable that Pine and Gilmore themselves now recognise this, in their 2011 revised edition of their 1999 work.

(3) Curiously, there is little in the Experience Economy model regarding differences in consumer expectations. Critics argue that certain market segments find little value in rigidly constructed performances. Florida (2002) identifies the growing importance of a *Creative Class* in Western economies. Well-paid, independent and culturally savvy, they are the drivers of economic development and the most desired target market in tourism. However, their tastes are discerning. They demand 'more active, authentic and participatory experiences, which they can have a hand in structuring' (Florida, 2002: 167). In contrast, they find mass tourism products 'irrelevant, insufficient or actually unattractive' (Florida, 2002: 36). A good example of how such consumers make or break a commercial experience was the attempt by Starbucks to penetrate the Australian coffee market. Starbucks, the epitome of the Experience Economy approach, opened chain stores following the same image, experience and business model that they had developed in the USA. However, they were unsuccessful, as Australia already had a well-developed espresso coffee culture, featuring small independent operators with a strong Italian image. A success in one market was a failure in another (Frost *et al.*, 2010).

(4) Commodification is at the heart of the Experience Economy. Pine and Gilmore argue that their approach allows businesses to charge higher prices. Indeed, it is fundamental to their theory that higher prices equate with better experiences and vice versa. Attempting to apply the model to cultural heritage, Richards (2001) posed some critical questions. Did cost really influence quality? Were experiences that were free not really experiences? What about experiences that were difficult to charge for, such as walking down the street in a historic European city? Was this somehow no longer valuable? We would like to take this a step further. Consider a spectacular natural landscape, such as the Grand Canyon. The US National Parks Service charges an entry fee of US$25 and provides a small amount of interpretation. For most visitors, the big experience is looking at the Grand Canyon, taking photos and perhaps sharing them with family or friends. Can that experience be enhanced through a scripted performance? Would that increase the satisfaction of visitors (justifying a higher charge)? We cannot see how that is desirable. The experience is so great that no value adding would improve it for most visitors.

(5) Design and scripting require a strong management structure. Examples given by Pine and Gilmore are of successful large corporations operating within the USA. Can that be transferred to the remote areas that are the settings for safaris and treks? A major part of the experience for explorer travellers is the interaction with indigenous and traditional peoples, even if they are only employed guides and support staff. Attempting to construct scripted performances for these cross-cultural encounters will be deeply unsatisfying.

These criticisms illustrate that, while the Experience Economy model is popular with suppliers, it is inappropriate for explorer travellers. Rather than a scripted and rehearsed performance, explorer travellers are seeking spontaneity, authenticity and personalised experiences. Even if they are on a packaged tour they still want those experiences. The tour may be chosen for access or expertise, but it cannot look too much like a mass tourism group tour.

Blurred Lines

In recent years, a number of operators have developed tours to cater for explorer travellers. These blur the lines between commercial tours and scientific and geographic exploration. While providing logistical organisation, access and expertise, they allow the tourists to perceive themselves as doing

Figure 10.4 The 2007 Arid Rivers Expedition along the Kallakoopah Creek in the Simpson Desert Regional Reserve in South Australia
Source: A. Harper, Australian Desert Expeditions.

something more than merely taking a holiday. For example, Australian Desert Expeditions (ADE) conducts scientific expeditions and surveys into the heart of Australia's Outback, in partnership with universities, government bodies and private research organisations, using pack-camels as transport, to maintain the lightest environmental footprint possible. ADE offers places to tourists to assist with the fieldwork, and in the process learn more about Australia's flora and fauna, as well as sustainability issues (Figure 10.4).

Four different types of tour/expedition blends are worth considering.

Voluntourism with wildlife

Voluntourism – holidays with a component of volunteering – are regarded as a rapidly growing area. They are particularly popular with young people and often have a focus on working to help endangered wildlife. For example, the Great Projects (www.greatprojects.com) offer a range of voluntourism opportunities in Africa and Asia. These include: habitat enrichment for orangutans in Sarawak; conservation and research on a private nature reserve in the Kalahari Desert in Botswana; cheetah conservation and release in Namibia; and working at a wildlife refuge in Zimbabwe. Another provider is the Orangutan Project (www.orangutan.org.au). With this, the tour raises funds for conservation and delivers supplies to an orangutan orphanage.

Self-drive safaris

The tourist drives a four-wheel-drive vehicle through Namibia. They set up camp and cook meals for themselves. The illusion is that they are on their own expedition. However, the organisers have developed the itinerary and provide back-up support. If the tourist runs into difficulties, they have a satellite phone and two-way radio to call in the nearby support vehicle (www.classicsafaricompany.com.au; www.selfdrivesafaris.com).

Mountain climbing expeditions

In the past, climbers of Mt Everest and similar charismatic peaks were elite adventurers. However, that changed in 1985 when Dick Bass, a 55-year-old Texas millionaire, was guided to the top of Everest. His achievement demonstrated two things. First, it was possible for an amateur climber to make it to the top if he had professional support. Second, there was a demand for such services. The result was a growth in tour companies offering such difficult climbs (Frost, 2004; Krakauer, 1997, and see later in this chapter). Nor are such operators confined to the Himalayas. Travel writer Dugald Jellie, for example, failed in his youth to climb Cotopaxi in Ecuador. Twenty years later, however, he was successful through joining a climb organised by Safari Tours (Jellie, 2012).

Exploration expeditions

A range of tour companies provide experiences either directly based on specific explorers or drawing on the more general myth. Their marketing images draw heavily on the Golden Age of Exploration. They are typically aimed at the high end of the tourism market, providing for wealthy Westerners (see, for example, www.shackletonandselous.com). Apart from these regular tours, there are special commemorative tours. An example of this is the re-enactment of Shackleton's 1912 Arctic voyage, staged by Tim Jarvis for the centenary (www.shackletonepic.com; Figure 10.5, see also Chapter 8). On this expedition, tourists could pay to participate, being accommodated on the support ship.

Challenges and Issues

Some interesting patterns are evident in these commercial tours. Four to consider are:

Figure 10.5 Tim Jarvis and Paul Larsen on board Alexandra Shackleton
Source: Ed Wardle/ Shackleton Epic.

Elitism and authenticity

There is a high premium on exclusivity in the advertising of tourist safaris and treks. These are not for the masses, rather they are packaged as proving that participants are members of an elite – a sort of modern-day explorers' club (for client perceptions of themselves, see Foster, 1986). Reinforcing this, they are marketed as the opposite of other products (which are presented in comparison as lame and lacking in status). Consider, for example, this newspaper advertisement for Orion Expedition Cruises, whose slogan – taken from Robert Frost – is *A Path Less Travelled*:

Wanted 100 People for an Expedition

Spiritual renewal is a big ask from a holiday. There's no resort that can provide it, no packaged vacation that can include it. Yet for those who look beyond 'a holiday', there's an endless source of that elusive replenishment.

Problem is, it's in remote and wild places.

An Orion expedition is not for everyone, clearly. Is it for you? ...

Few (phew) have the will to journey off the beaten track to such places (otherwise there might one day be highways).

But if you're blessed with that will, and the wherewithal, consider the Orion way: aboard a 5-star expedition vessel that has been rigorously engineered and is seriously equipped to take a path less travelled. ...

No roads get hacked, no rainforest sacked to accommodate Orion guests. (Orion Expedition Cruises, 2012)

Such advertising directly links the structure and meanings of these expeditions to authenticity. As argued by Cohen (2004) and Wang (1999), many modern tourists view authenticity in terms of how it makes them feel. They are attracted by experiences that they see as real and as enhancing their satisfaction and identity. Accordingly, they value tourism products that promote the sense that they are elite, avoiding the masses and enjoying special service and status. Exploring and adventure fit exactly with this concept of the appeal of existential authenticity (see, for example, Arnould & Price, 1993; Gyimóthy & Mykletun, 2004).

Managing the attraction of risk

The growth of adventure tourism has sparked great interest in how risk may be managed, particularly how it may be minimised on commercial tours (Frost, 2004; Wilks *et al.*, 2006). However, it is the thrill of risk that attracts many explorer travellers to undertake tours. Rather than wanting to be risk minimisers – having completely safe experiences – many gain their satisfaction from the danger, adrenaline rush and challenge of achieving something in the face of major obstacles and extreme environmental conditions. Accordingly, safari-style tours and treks are paradoxical, with many participants seeking the benefits of risk. Indeed, they are making a personal calculation that 'the benefits of risk-taking are far more likely than the risk of loss' (Ryan, 2003: 57). This trade-off might entail accepting risks in order to gain the benefits of greater insights and skill development (Walle, 1997).

Critical to the management of risk is the matching of expectations and capabilities. There is a paradox in the marketing of adventure and exploration tours and the preparation and skillsets of participants. As Guy Cotter, operator of the climbing company Adventure Consultants, explained:

> For someone to come to Mount Everest with us, we want to be sure there is a good chance of them succeeding, so we look for a strong climbing background followed by [training through] gradual ascents to higher altitudes. This way we know the clients have the experience to succeed and won't have wasted money trying to do something they have fantasised about, but are totally incapable of doing. (quoted in Upe, 2001: 14)

The roles of tour guides

Many explorers and adventurers lack resources and accordingly work as tour guides. This provides them with opportunities, but also leads to tensions and blurred responsibilities. These issues were particularly apparent in the 1996 deaths resulting from two tour companies climbing Mount Everest.

Into Thin Air (Jon Krakauer, 1997)

Mountaineering journalist Jon Krakauer accompanied a climb of Mt Everest staged by the commercial company Adventure Consultants. This enterprise had been formed by New Zealand climbers Rob Hall and Gary Ball. As one of their colleagues explained to Krakauer, they had gained sponsorship for previous climbs, but:

> To continue receiving sponsorship from companies, a climber has to keep upping the ante. The next climb has to be harder and more spectacular than the last ... so they decided to switch direction and get into high-altitude guiding. When you're guiding you don't get to do the climbs you necessarily most want to do; the challenge comes from getting clients up and down ... But it's a more sustainable career than endlessly chasing after sponsorships. There's a limitless supply of clients out there if you offer them a good product. (quoted in Krakauer, 1997: 35)

Adventure Consultants became the leading company in escorted ascents, between 1990 and 1995 guiding 39 climbers to the summit. The price by 1996 was US$65,000 per head (Krakauer, 1997: 37). Also climbing Mt Everest on the same day was another commercial company, Mountain Madness. Both groups got into trouble after they reached the summit. Descending late in the day, they were caught by a blizzard and three guides and two clients died.

Krakauer identified a number of problems with the relationships between tour companies, guides and clients. For Scott Fischer's Mountain Madness company, there were difficulties with head guide Anatoli Boukreev. A highly experienced climber, he was paid a US$25,000 fee, in contrast to the usual US$10,000–15,000. However, the problem was:

> Boukreev came from a tough, proud, hardscrabble climbing culture that did not believe in coddling the weak ... He was quite outspoken in his belief that it was a mistake for guides to pamper their clients. 'If client cannot climb Everest without big help from guide,' Boukreev told me, 'this client should not be on Everest' ... Boukreev's refusal or inability to play the role of a conventional guide in the Western tradition exasperated Fischer. It also forced him and [third guide] Beidleman to shoulder a

disproportionate share of the caretaker duties for their group. (Krakauer, 1997: 155–156)

Furthermore, Boukreev was an advocate of climbing Everest without bottled oxygen. In the macho world of mountaineering, this was seen as the ultimate challenge, the mark of the elite climber. However, while Boukreev had twice before climbed Everest without oxygen, on this trip he was being paid to guide tourists. Krakauer noted the lack of oxygen:

> I was surprised Fischer had given them permission to guide the peak without it, which didn't seem to be in their clients' best interest. I was also surprised to see that Boukreev didn't have a backpack … containing rope, first-aid supplies, crevasse-rescue gear, extra clothing and other items necessary to assist clients in the event of an emergency … because Boukreev wasn't breathing supplemental oxygen, he had apparently decided to strip his load down to the bare minimum to gain every possible advantage in the appallingly thin air. (Krakauer, 1997: 186–187)

Severely disadvantaged by his lack of oxygen, guide Boukreev was the first to depart the summit and climb down. The clients and other guides were far slower. The chief difficulty with Everest was that the climbers needed to turn around by 2pm, whether or not they had reached the summit. Enforcement of this fell apart. Only six of 34 had reached the top by that time and the others kept going. As guide Neal Beidleman explained:

> It was supposed to be Scott's job to turn clients around … I'd told him that as the third guide, I didn't feel comfortable telling clients who'd paid sixty-five thousand dollars that they had to go down. So Scott agreed that would be his responsibility. But for whatever reason, it didn't happen. (quoted in Krakauer, 1997: 208)

For Adventure Consultants, the problem was waiting too long for the slowest climber to reach the top. Owner Rob Hall arrived at the summit just after 2pm. He then waited for client Doug Hansen. A year earlier, Hall had turned Hansen around at 2.30pm. He had worked hard to convince Hansen to make another trip. Having invested such emotional labour in Hansen, it is likely that Hall felt duty bound to ensure that this time Hansen succeeded. Hansen did not arrive until 4.30pm. Another guide, Andy Harris, returned to the summit to help them. All three died.

An interesting perspective on the Mt Everest tragedy was provided by Graham, the Outback explorer we interviewed. He had read everything on the disaster he could lay his hands on and made direct comparisons to his own experiences:

> I believe expeditions fail before they leave ... on that expedition you could see the progressive mistakes being made ... the mistakes built on themselves as they got up the mountain and at any number of various points, the whole thing could have been avoided ... and the thing about the way I do it ... is all this planning that goes in beforehand means that if you have too much adventure, and this is one of my mottos, too much adventure means there's been too little planning ... you would have to question the abilities and capabilities of the people who led the expedition. They were fabulous climbers, but that doesn't mean they were going to be fabulous guides ... I believe all these tourist type experiences, even the adventure ones, should always proceed at the pace and capability of the weakest person in the party.

Working with host communities

As they operate in remote and wild destinations, tourist safaris and treks interact with indigenous and traditional communities. For explorer travellers, this is part of the appeal. Unlike the more mundane mass tourism products, they are seeking to get up close and personal with people and cultures quite different from their own. The essence here is the exploration of these exotic cultures. Again the issue of authenticity is paramount and it is very much the existential authenticity of the visitors feeling they are having a real and special experience. In the imagining of modern tourism, this cannot be achieved through staged performances for large groups, but rather through incidental encounters by small groups.

The impact of tours on local economies is highly contested. Zurick (1995) argues that the paradox of adventure travel is that the more a remote place is explored, the more spoiled and unattractive it becomes. In contrast, Krakauer rejects such views:

> It seems more than a little patronizing for Westerners to lament the loss of the good old days when life in the Khumbu [region] was so much simpler and more picturesque. Most of the people who lived in this rugged country seem to have no desire to be severed from the modern world ... The last thing Sherpas want is to be preserved as specimens in an anthropological museum. (Krakauer, 1997: 48)

Krakauer argues that the region is already tied to the world economy and that tours have a major economic impact:

> For better and worse, over the past two decades the economy and culture of the Khumbu has become increasingly and irrevocably tied to the seasonal influx of trekkers and climbers, some 15,000 of whom visit the region annually. Sherpas who learn technical climbing skills and work high on the peaks – especially those who have summitted Everest – enjoy great esteem in their communities ... Despite the great hazards, there is stiff competition among Sherpas for the twelve to eighteen staff positions on the typical Everest expedition. The most sought-after [are] ... for skilled climbing Sherpas, who can expect to earn $1,400 to $2,500 for two months [compared to] ... an annual per capita income of around $160. (Krakauer, 1997: 47)

The involvement of indigenous and traditional peoples in nature-based and adventure tourism is changing. In the past, there was a tendency to ignore them, in some cases even to force them out of national parks and protected areas. Recent developments have included a shift from mere menial labour to greater participation in the management and ownership of protected areas, accommodation and tours, particularly safaris and treks (Hall, 2000; Nepal, 2000; Zeppel, 2009).

The ideal for the explorer traveller is taking part in safaris or treks in which indigenous and traditional peoples are the guides. This allows the visitor to ask questions and discuss issues with a local person. An example of this is Harry Nanya Tours, an indigenous owned and operated small tour company that takes visitors to Lake Mungo in Outback Australia (Figure 10.6). Lake Mungo is a World Heritage Site, famous for archaeological excavations of ritual burials from over 20,000 years ago – making these examples of the oldest religious ceremonies in the world. In a remote location, distance and the associated cost filters out tourists, so that those who journey there tend to have a very strong interest in culture and archaeology. A key part of the appeal of the tour is that the guides have also worked on the archaeological digs and are able to discuss that as well as indigenous culture.

Some tours offer the opportunity for the visitors to *immerse* themselves in an indigenous or traditional culture. An example of this is Tiwi Tours (Venbrux, 2000). The Tiwi Islands (also known as Bathurst and Melville Islands) are 80 kilometres north of Darwin, Australia. The local Aboriginal community established Tiwi Tours to attract tourists and generate employment. Those drawn to the tour are interested in Aboriginal culture and in having an adventurous experience in an isolated part of the world

Figure 10.6 Guided tour at Lake Mungo National Park, Australia
Source: W. Frost.

where few tourists have been. They are mainly young independent travellers, who take the tour as a way of going beyond the well-worn tourist attractions of Darwin.

Participants in the tour are immersed in the community and engage in day-to-day activities. Women, for example, accompany two Aboriginal women who collect traditional food along the beach and in the bush. Going beyond observing or asking about these practices, the travellers join in, being shown how to chip shellfish off rocks, dig for roots and chop open logs to search for small game.

As traditional communities become more involved in tourism, there is an increasing issue of inequities and tensions within these societies. This is particularly so when the benefits are not evenly distributed. We conclude this chapter with an example of such a problem.

The Song of the Dodo (David Quammen, 1996)

Science writer David Quammen journeys to Madagascar to see a rare lemur. He needs a guide. His hotel recommends Bedo, a young Malagasy boy, as the best:

I had already spent an afternoon in the reserve with another guide, a precociously cynical youngster who presumed that I'd tip him well for manhandling the boa constrictors and chameleons into photogenic poses. Bedo seemed promisingly different. For one thing, he would gladly take me through Analamazaotra [forest reserve] by night, when the nocturnal lemurs were at large … It was an overcast night with hardly a trace of moon. Bedo set a brisk pace … His attunement to the ecosystem seemed preternatural. To me he paid little notice, as long as I kept pace and appreciated the animals after he spotted them. He was no cynic, pandering to the predictable tourist impulses, angling for tips. He was a naturalist, inexhaustibly intrigued by his subject and tolerant of the twit who chose to tag along. (Quammen, 1996: 508–509)

Later Quammen learns that Bedo is famous for guiding world-class scientific researchers and photographers. Some of the scientists want Bedo to continue his schooling, but money and Western attention are more seductive. Quammen reflects that he became 'another of the foreigners who fawned on him, paid him well, and made his work as a guide seem more promising than further education' (Quammen, 1996: 523). Bedo's talent leads to tragedy and he is murdered:

Why was he murdered? Because, in becoming the darling of foreign tourists, in earning so much fast easy money as a guide, he had provoked envy. Malagasy culture does not tolerate upstarts. Bedo spent his money on a punk haircut, a radio, flashy clothes. He had gotten abrasive and highfalutin. Some people hated him for it. (Quammen, 1996: 545)

Part 4
The Future

11 Destination Mars

What is the future for exploration in an era where the peripheries are already mapped or settled? Even in the early years of the last century, Joseph Conrad writes of his disappointment that explorers and discovers are becoming obsolete, given that 'there is nothing new left now and but very little of what may still be called obscure' (Conrad, 1928: 88). Yet the adventurer Benedict Allen is clear that exploration is still possible in the 21st century:

> The truth is that we have named only one and a half million of this planet's species, and there may be more than 10 million – and that's not including the largest group of all, bacteria. We have studied only 5 per cent of the species we do know. We have scarcely mapped the ocean floors, and know even less about ourselves; you could say we fully understand the workings of only 10 per cent of our brains. (Allen, 2002: xii)

Interestingly, he does not mention the current growth in space tourism, which is viewed as a major future direction in travel (Cater, 2010; Crouch et al., 2009; Laing & Crouch, 2004; Reddy et al., 2012), nor the unexplored solar system. It is likely that the explorer traveller in years to come will be bound for interstellar space and even Mars (Murphy, 2010; Walter, 1999; Zubrin, 1996). President George W. Bush, in a speech at the University of Tennessee on 2 February 1990, compared space exploration with the opening up of the American frontier: 'And just as Jefferson sent Lewis and Clark to open the continent, our commitment to the Moon/Mars initiative will open the Universe'. This is also the dream of many of the explorer travellers we have interviewed. They mostly expressed an interest in space travel, although some had reservations about the way this might be carried out and how much autonomy they would have over the experience. Many had read or watched TV shows or films about science fiction, both as children and adults (Laing & Crouch, 2009a) and saw outer space as the logical next step in exploration.

In this chapter, we contrast how expeditions to Mars have been imagined in the past with current scientific expectations of what space tours will be like in the future, and how one might influence the other. Central to any journey to Mars is the paradox of the normally independent traveller confronted with an experience that is likely to be highly regulated and controlled by others, as well as heavily reliant on technology. This may be at odds with the traveller's fantasy of space tourism as an individualistic pursuit, requiring specialised skills and initiative possessed by those with *the right stuff*.

The Race for Space

The backdrop to space exploration to date has largely been the desire for governments to develop space programmes for reasons of national prestige. The space race between the United States and the Soviet Union started with the shock of Sputnik 1, the first artificial satellite, in 1957, followed by the sending of animals into space and Yuri Gagarin's orbit of the Earth in 1961, making him the first human being in space. Soviet domination of space was seen both as a threat and a blow to American pride. John F. Kennedy is said to have remarked bitterly after the Soviet recovery of two dogs from orbit in August 1960: 'The first living creatures to orbit the earth in space and return were dogs named Strelka and Belka, not Rover or Fido' (quoted in Gadney, 1983: 129–130).

The Americans were swift to reply in kind. The Mercury Program was set up in 1959 (Figure 11.1) and its seven astronauts – Alan Shepard, John Glenn, Scott Carpenter, Gus Grissom, Gordon Cooper, Deke Slayton and Wally Schirra – became household names and featured in *Life* magazine. All former test pilots, their status and lifestyle were brought to life in Tom Wolfe's *The Right Stuff* (1979). Children held them up as heroes and they provided the names for International Rescue's Tracy brothers in the 1960s classic *The Thunderbirds*. These seven were the 'elite who had the capacity to bring tears to men's eyes' (Wolfe, 1979: 24). Something of this mythic glamour still attaches itself to the astronaut, and perhaps to the space traveller of the future.

Some of President John F. Kennedy's classic and most inspirational speeches related to space travel, including the one he delivered to Congress on 25 May 1961. The response was so quiet that Kennedy later told his adviser on the drive back to the White House that he felt the response was 'less than enthusiastic' (Sorensen, 1965: 526). Perhaps some were silenced by a sense of the historical import of the speech, while others were simply stunned at the ambitious task Kennedy had set the nation with his call to arms:

> I believe this nation should commit itself to achieving the goal, before this decade is out, of landing a man on the Moon and returning him safely to the Earth. No single space project in this period will be more

impressive to mankind, or more important for the long-range exploration of space; and none will be so difficult or expensive to accomplish ... In a very real sense, it will not be one man going to the Moon – if we make this judgement affirmatively, it will be an entire nation. For all of us must work to put him there.

Despite Kennedy's initial concern at Congress' subdued reaction to the speech, they approved a massive increase in NASA's budget, which burgeoned from approximately US$0.5 billion in 1960 to US$5 billion in 1965 (Shipman, 1987). Yet sadly the focus on landing a human being on the moon was arguably one of the reasons for the US space program running out of steam shortly after this goal was met. Kennedy's assassination led to successive Presidents (Johnson and Nixon) who saw the achievement of the first landing as the logical *conclusion* to the Apollo program, rather than the *start* of a golden era of space exploration. The public largely shared this view. Once the Apollo 11 crew returned from their history-making lunar flight, public interest noticeably waned. By the flight of Apollo 13, networks no longer automatically showed mission

Figure 11.1 Mercury Capsule 15B. Freedom 7 *II*. On display in the Smithsonian National Air and Space Museum, Washington, DC
Source: W. Frost.

footage on primetime television, that is, until the Apollo 13 crew's lives were put in jeopardy and attention was caught by the drama. It has been argued that the American public are only interested in *exceptional* space achievements, perhaps influenced by the science fiction they read or watch on TV (James, 1999). As Shirley (1998: 5) observed: 'popular culture fostered an interest in space, but the real world rarely delivered the goods'. Where public opinion went, political opinion was not far behind and was one step ahead in some cases. Congress slashed the NASA budget to its lowest level in 10 years, and the last three Apollo flights, 18, 19 and 20, were scrubbed. Gene Cernan and Harrison Schmitt became the last two human beings to walk on the moon in December 1972, and Project Apollo was now to all intents and purposes over. It is now over 40 years since anyone has gone to the Moon.

The New Frontier: Mars

From the Apollo–Soyuz docking in 1975 until the first Space Shuttle flight in 1981, the US presence in space ground to a halt. Since then, we have seen the development of the International Space Station (ISS) and a more co-operative approach to space exploration. While missions back to the Moon and on to Mars have been mooted, timetables have been put back again and again and the Shuttle (Figures 11.2 and 11.3) has been retired, leaving Russian

Figure 11.2 Space Shuttle, on display in the Smithsonian National Air and Space Museum, Washington, DC
Source: W. Frost.

Figure 11.3 Space Shuttle Endeavour on the launch pad, Kennedy Space Centre, Florida
Source: J. Laing.

Soyuz rockets as the only means of transport to the ISS. Proponents of Mars exploration, such as Robert Zubrin and Apollo astronaut Buzz Aldrin, are actively lobbying governments and arguing the case for continuing space exploration beyond near-earth orbit. Zubrin established the Mars Society, with chapters around the globe, to act as advocates for exploring Mars, and their activities include conducting research in analogue (Mars-like) places on Earth such as Utah, the Canadian Arctic and the South Australian desert (Murphy, 2010). In 2001 Jennifer was involved in an expedition through the Australian Outback to find a suitable location for the Australian Mars Analogue Research Station (MARS-OZ) and spent time in 2003 living on the Mars Desert Research Station (MDRS) in Utah (Figure 11.4).

The delay in a Mars mission has saddened and disappointed some of our interviewees. Graham wishes he had been born in another era: 'Because

Figure 11.4 The Mars Desert Research Station, Utah
Source: J. Laing.

terrestrial exploration's finished and the real period of man's next exploration [outer space] hasn't started'. Emily is concerned that she won't see a Mars landing as the timelines get pushed out: 'I don't think we're going to be colonising Mars for decades from now ... Some of us are getting to the age where we'd like to see things speeded up. So we can know that there's a Mars colony or a space colony, a Moon colony, you know, in our lifetimes. Because we're not getting any younger'. Setting foot on Mars is seen as the inevitable next step for explorer travellers. As Geoff observes:

> I think if you had a spaceship that was put together and it needed a guy to go to Mars or the Moon or something, a commercial expedition, I don't care if it's held together with masking tape, I'd want to be on it. It would be, for me, amazing, because that in a sense is actually what it's all about, that is the new Shackleton, the new Christopher Columbus. This is the future. This is about pushing the human race forward. That's what exploration's about.

The most recent pronouncement by NASA administrator Charles Bolden is more optimistic about a Mars mission: 'A human mission to Mars is today the ultimate destination in our solar system for humanity, and it is a priority for NASA. Our entire exploration program is aligned to support this goal' (Prigg, 2013). Bolden announced a proposed NASA mission by 2025 to capture and relocate an asteroid to Earth's orbit, allowing astronauts to land on the surface and take samples for analysis. This could then be used as a launching pad for a mission to Mars. NASA has been criticised for not leaping straight into a Mars program. However, as Bolden explains: 'We don't have the technological capacity right now' (quoted in Roberts, 2013). Other research is planned to prepare the way for long-duration space flight and exploration. In 2015 a US astronaut will live for a year on the ISS, to examine the effects of zero gravity on the human body.

The Americans may face competition from other space programmes, notably China, which has announced its own plans for sending humans to Mars by 2040 (Roberts, 2013). There might be some impetus for joining forces, along the lines of the ISS project. Ultimately, if we are to send human beings to Mars and return them safely to Earth, it will take a tremendous collective effort and will, perhaps involving more than one nation, and a huge funding commitment, spanning many years. Without the impetus of a firm goal and timetable, such as Kennedy set in 1961 for the moon landings, such a feat is unlikely to be realised. But if past mistakes are to be avoided, we should not place the emphasis purely on getting to Mars, if we want to establish a long-term presence in space.

Once Upon a Time, in a Galaxy Far, Far Away

Human beings have been dreaming about space travel for centuries and it has long been a popular subject for literature (Ashley, 2011). Its roots can perhaps be traced back to imaginary travel narratives, such as the *Epic of Gilgamesh* and the *Odyssey*, as well as utopian literature, which focused on fantastical travel and the discovery of an unexplored haven (Ashley, 2011; Hulme & Youngs, 2002; James, 1999; Laing & Frost, 2012). Some of this genre has a satirical edge, such as Lucian of Samosata's *Vera Historia* (*True History*) (AD 160), which depicts flight to the Moon using the wings of an eagle and a vulture, and Thomas More's *Utopia* (1516). The concept of a utopia, an idyllic place or society, can be found in early Greek literature, with the play *Ornithes* (*The Birds*) by Aristophanes (*c.* 414 BC). (Ashley, 2011). Utopian themes are still popular within fiction, such as Shangri-La in James Hilton's *Lost Horizon* (1933), H.G. Wells' *A Modern Utopia* (1908) and Aldous Huxley's *Island* (1962).

The father of science fiction is Jules Verne, whose novel *From the Earth to the Moon* (1865) made audiences fantasise about what it would be like to travel to outer space and how they might get there. Around the turn of the century, Mars captured the public imagination, thanks to a mistaken translation of the work of an Italian astronomer, Giovanni Schiaparelli, who described a network of *canali* on Mars, the Italian word for channel.

This was interpreted as canals, the work of an alien being, rather than a natural phenomenon (Murphy, 2010). Scientists like Percival Lowell argued that Mars was inhabited (Dick, 1998). This led to a spate of Martian-themed popular novels such as H.G. Wells' *War of the Worlds* (1898), a tale of a Martian invasion, and Edgar Rice Burroughs' *A Princess of Mars* (1912). In the latter, the hero, American Civil War veteran John Carter, uses the power of the human mind 'to move the body to the stars' (James, 1999: 266), a novel way of explaining away the lack of any ship to Mars. There he discovers a constant state of war and a people who have evolved the ruthlessness necessary to survive in a world of dwindling resources. These narratives fed the public appetite for exploring the possibility of life on other planets, but also popularised the idea of aliens as something to be feared (Ashley, 2011).

The War of the Worlds was a notable influence on two fronts, firstly for the impact it had as a novel, and then for the mass panic it inspired when broadcast as a radio play read by Orson Welles in 1938. Even on the cusp of a World War, Americans were more concerned about and willing to believe in an extraterrestrial threat. Its opening words are as chilling today as when they were first written, even if we no longer fear a Martian invasion:

> No one would have believed, in the last years of the nineteenth century, that human affairs were being watched keenly and closely by intelligences greater than man's and yet as mortal as his own; that as men busied themselves about their affairs they were scrutinized and studied, perhaps almost as narrowly as a man with a microscope might scrutinize the transient creatures that swarm and multiply in a drop of water. (Wells, 1898: 1)

Wells goes on to describe the consequences of this scrutiny; how 'across the gulf of space, minds that are to our minds as ours are to the beasts that perish, intellects vast and cool and unsympathetic, regarded this earth with envious eyes, and slowly and surely drew their plans against us' (Wells, 1898: 1). Douglas Adams pays homage to this in the beginning of *The Hitchhikers' Guide to the Galaxy* (1979), where he writes of the Vogon spaceships who are about to obliterate the Earth to make way for a 'hyperspatial express route': 'They soared with ease ... biding their time, grouping, preparing ... The planet beneath them was almost perfectly oblivious of their presence, which is how they wanted it for the moment' (Adams, 1979: 20). The alien as a cruel and heartless oppressor continues to be a popular theme for science fiction, spawning such memorable villains as the Daleks and Cybermen in *Doctor Who* and the Borg in *Star Trek* (Ashley, 2011).

Modern science fiction classics involving explorer journeys include *Beyond the Silent Planet* (C.S. Lewis, 1938), *Starship Troopers* (Robert Heinlein, 1959), *Dune* (Frank Herbert, 1965), *Rendezvous with Rama* (Arthur C. Clarke, 1973) *The Forever War* (Joe Haldane, 1974) and *The Martian Race* (Gregory Benford, 1999). These books often introduce crises into space travel, such as solar flares, meteorite showers or encounters with aliens, to ratchet up the tension and overcome the tedium of a long journey (James, 1999). While some science fiction is deliberately fantastical, other authors take great pains to ensure scientific accuracy, particularly if they write about places such as the Moon and Mars, about which we know a great deal and where the body of knowledge is increasing, thanks to robotic missions (Murphy, 2010; Shirley, 1998). Despite these real-life advances in understanding, we paradoxically crave more imaginative depictions of the universe in fiction, to maintain a mysterious realm into which we can escape (James, 1999).

These books fulfil another role 'as one element of the propaganda drive for space travel' (James, 1999: 252), highlighting the inevitability of space travel in terms of human destiny. Conversely, this literature is a reaction to real-life events. President George Bush Sr.'s announcement in 1990 about the goal of a Mars mission led to 'a flurry of Mars novels in the 1990s' (James, 1999: 257), such as Kim Stanley Robinson's *Mars* trilogy (1992–1996). Science fiction has heightened expectations that space travel would happen sooner rather than later. The influence of Kubrick's film *2001: A Space Odyssey* (1969), based on an Arthur C. Clarke novel *The Sentinel* (1951), was mentioned by Emily, who thought it was a portent of the ubiquity of space tourism by the turn of the 21st century:

> For those of us who grew up reading science-fiction, [space travel] was only natural ... And so I think a lot of us were disappointed when the year 2001 came and we're still not up there, we're just in low-Earth orbit.

Science fiction became a staple of popular culture in the 20th century. Many comics have utilised space themes, with heroes such as *Flash Gordon* (1934–2003), DC Comics' *Green Lantern* (1940–) and Marvel's *Guardians of the Galaxy* (1969, 2008). Toys were created to cash in on the interest in space travel. Jennifer's brother built a replica model of the Saturn V rocket and she had an astronaut Barbie doll, while Warwick owned a model of the Thunderbirds. The Roswell incident in 1947, where an alien spacecraft was said to have crash-landed in the New Mexico desert, coincided with a spate of UFO sightings and, like Area 51, became 'part of American folklore ... and testimony to American scepticism of government' (Dick, 1998: 166). This led to the universal depictions of aliens as small, with a large cranium and huge bug-eyes. In the 1950s, *Collier's* magazine published a series of illustrated articles on space which reached a circulation of four million, and Walt Disney developed a space-themed segment of his Disneyland theme park known as Tomorrowland (Laing & Crouch, 2004; Smith, 2000).

Space travel on television and in film also burgeoned in the modern era, particularly as technological developments like CGI have improved special effects and made production more cost-effective. The *Star Trek* (1966–) franchise has enduring appeal (Kozinets, 2001), as does *Doctor Who*, which is celebrating its 50th anniversary in 2013. Both of us grew up watching these series, along with *Lost in Space* (1965–1968) (see Chapter 7), and it fostered our lifelong interest in science fiction. The novelty of *Doctor Who*'s plot involved time travel in a spaceship known as the TARDIS (Figure 11.5), which allowed the Doctor to transcend long space journeys and the 'challenge of Einsteinian physics' (James, 1999: 261). Cartoons based in outer space such as *Space Ghost* (1966–1968), Warner Brothers' Marvin the Martian, *Space Ace* and *Prince Planet* and animated movies like *Disney's Ducktales: Space Invaders* (1990) and *Treasure Planet* (2002), give children an introduction to space as an environment

Figure 11.5 TARDIS on display in the Dr Who Up Close Exhibition, Cardiff, Wales
Source: J. Laing.

where the extraordinary (and aliens) are commonplace. For adults, TV spoofs on space travel, such as *Red Dwarf* (1988–) and *The Hitchhiker's Guide to the Galaxy* (book 1979, TV series 1981, film 2005), allow us to laugh at aliens and neutralise our fears of space as a cold and forbidding place. Even the *Muppets* got in on the act, with their regular 'Pigs in Space' skit.

Numerous space-themed films have been and still continue to be made. One of the first movies ever made was *Le Voyage Dans La Lune* (*A Trip to the Moon*) (1902) by Georges Méliès, a feat dramatised in the recent film *Hugo* (2011). Audiences in the 1930s thrilled to the first Flash Gordon escapade, *Rocketship* (1936) and *Buck Rogers* (1939). In the 1950s, many B-grade movies, now firmly in the cult camp, were associated with science fiction, such as *The Blob* (1958) and Ed Wood's *Plan 9 from Outer Space* (1959). Both of us remember spending Christmas days watching *Santa Claus Conquers the Martians* (1964), on TV – a desperate attempt to combine two themes that were apparently guaranteed to appeal to children. Of similar vintage were *The Three Stooges in Orbit* (1962) and *Robinson Crusoe on Mars* (1964) – which despite its name had pretensions toward being a serious drama. True cinematic classics were also produced during the Cold War such as *Forbidden Planet* (1956), a galactic version of Shakespeare's *The Tempest*, introducing Robby the Robot, and *The Day The Earth Stood Still* (1951), which introduced the notion of aliens who are wiser and more peaceful than we are on Earth and thus must teach us a lesson before it is too late. In 2008, the version starring Keanu Reeves changes this to an environmental message. Tim Burton satirises this trope of the *noble alien* in *Mars Attacks!* (1996), where the Martians pretend to negotiate but in fact just want to let rip with laser guns. Like the Martians in *The War of the Worlds*, these aliens are defeated, but not by weapons. The fatal culprit was bacteria in Wells' novel and a yodelling version of the song *Indian Love Call* in Burton's film.

The 1960s saw sex in space, epitomised by Jane Fonda with her boots, tight jumpsuit and tousled hair in *Barbarella* (1968). This was juxtaposed with the eternally puzzling and poetically beautiful *2001: A Space Odyssey* (1969), complete with a classical music soundtrack, which restored a grandeur and seriousness to stories of space flight. The backdrop to these films was the Moon landings, but interestingly both set their stories into the future (the 41st century in the case of *Barbarella*), when they can indulge in fantastical speculation, rather than a realistic depiction of current space travel. Space tourism was highlighted, with tourists in *2001* flying to space stations en route to the Moon in Pan-Am modified shuttle spacecraft, and regular tourist flights to the Moon and Mars in *Moon Two Zero* (1969). The wonderful *Five Million Years to Earth* (aka *Quatermass and the Pit*, 1968) pre-empted *2001* with its storyline that in the past Martians genetically modified apes to create humans. It also, as in all the Quatermass films, posited the concept of there being a British Space Programme.

Space travel still holds its fascination in recent times and fantasy still remains the dominant story-arc, including the original *Star Wars* trilogy (1977–1983) and their more recent sequels (1999–2005). Explorer narratives and mythic constructs are common. *Star Trek*, in particular, uses the imagery of discovering far-flung frontiers throughout the universe (James, 1999; Kozinets, 2001). The mission of the Starship Enterprise is to 'explore new worlds, to seek out new life and new civilizations, to boldly go where no one has gone before'. Director J.J. Abrams plays with this concept in the latest film in the franchise, *Star Trek Into Darkness* (2013), where Scott (Simon Pegg) protests against the loading of torpedoes on board the Enterprise: 'I thought we were explorers!' These films were an influence on our interviewee Leo, who was training to become a space tourist: 'Of course I was born in the age of *Star Wars* so ... that does it too'. Despite robotic missions which have failed to find life on Mars thus far, and scientific consensus that the most likely form of life would be microbial (Cockell, 2007; Murphy, 2010; Walter, 1999), modern movies persist with the idea of aliens on Mars, including *Total Recall* (1990), *Red Planet* (2000) and *John Carpenter's Ghosts of Mars* (2001). In a similar vein, Roswell, with its UFO festival and museum (Figure 11.6) about the alien landing in 1947, is a popular destination for tourists (Paradis, 2002), and in 1997 thousands gathered in the small US town to celebrate the 50th anniversary; ironically at the

Figure 11.6 UFO Museum, Roswell, New Mexico
Source: W. Frost.

same time as the Pathfinder landing on Mars (Dick, 1998). It seems we still yearn for the alien encounter, however remote it seems in reality.

Now that we have ventured away from Earth, we also have numerous autobiographical accounts of what it is like to travel in space. A number of astronauts have written accounts of their journeys to the Moon or about life on a space station (Laing & Frost, 2012). The seven Mercury astronauts wrote their version of the early space flights in *Into Orbit* (Glenn *et al.*, 1962), described on the dust jacket as 'thrilling first-hand accounts by the men who took their lives in their hands and dared to leave the earth behind them'. Some of the most interesting are less concerned about maintaining their status as a hero and more about the challenge of space travel. Buzz Aldrin's book, *Return to Earth* (1973), is a searingly honest account of his difficulties when he returned home from the first Moon landing in 1969. What was left to achieve when one had participated in what was said to be the greatest feat of humankind to date? For such a goal-driven individual, there was no challenge left. Perhaps this is part of the reason why Aldrin has taken up the mantle of space advocacy since then so vocally and relentlessly.

Space Tourism

While settlement in space might be a long-term goal, space tourism is inching closer to becoming reality. While individual tourists such as Dennis Tito and Mark Shuttleworth have paid vast sums of money (reputedly US$20 million) to fly to and spend time on the ISS (Laing & Crouch, 2004), this is essentially a niche activity, only available to the elite. There have been moves to offer space tourism to the public on a wider scale, albeit to those individuals with a reasonably healthy chequebook. In 2004 the Ansari X-Prize, worth US$10 million, offered to the first non-government organisation that could fly three people into space and back (defined as 100 kilometres vertically from the surface of the Earth), was awarded to Burt Rutan's Scaled Composites, which developed *SpaceShip One* and an airborne launcher, the *White Knight*. The aim of the prize had been to stimulate the private sector to kick-start the space tourism industry. Entrepreneur Richard Branson then bought the technology for Virgin Galactic, and harbours a desire to fly in space one day on one of his Virgin Galactic vehicles. He writes: 'Space is the ultimate frontier – I'd love to be aboard our first flight' (Branson, 2004: 3). He also plans to take his family along, which is a shrewd strategy to allay public fears in the fledgling business and demonstrates his personal commitment to space flight. Branson is not the only billionaire interested in space tourism, but is perhaps the most high profile. Microsoft co-founder,

Paul Allen, is building rockets that could one day take people into orbit, while PayPal founder Elon Musk established a rocket company called SpaceX which has contracts with NASA (Chang, 2011).

Another clever move by Branson is to highlight celebrity endorsement of his technology. Among over 600 individuals reputed to be signed up for a Virgin Galactic flight (a snip at US$200,000) are said to be Brad Pitt and Angelina Jolie, Princess Beatrice of York (whose boyfriend Dave Clark works for Branson) and scientist Stephen Hawking (McKie, 2012). The Virgin Galactic website refers to this as 'one of the most exclusive clubs in the world' (Virgin Galactic, 2013). Two Australians also have a booking – Alan Finkel, founder of Axon Instruments and Chancellor of Monash University and Wilson da Silva, the editor of *Cosmos* magazine. Finkel's enthusiasm for space is described as: 'a gut-level desire for me. I am frankly excited about being up in that void and seeing the black sky punctuated by non-sparkling stars. I can't wait to feel the slow motion of gravity'. He argues that the price tag offered by Branson's company is 'an accessible amount. People spend more than that on cars'. It covers not just a seven-minute experience of weightlessness in space, but 'three years of anticipation [actually seven so far!], then six days of training … The actual flight will be two to three hours of circling up in the mother ship and then some tense minutes of being launched up into space'. Da Silva is equally excited to venture into space, but makes a telling point about the space tourism experience. Virgin Galactic asked whether they should provide a meal and have a flight attendant on board. Finkel was aghast. As da Silva observes: 'Space flights are supposed to be "only the brave" and a hostie [flight attendant] asking if you want a Diet Coke would ruin that' (quoted in May, 2006: 18).

Departures and arrivals will take place at Spaceport America in Las Cruces, New Mexico. While construction of the spaceport has yet to be completed, interested visitors can take a bus tour of the site (Spaceport America, 2013). New Mexico also hit the headlines in 2012 for hosting the 24-mile jump by Felix Baumgartner from a balloon. With the space connections of Roswell and the Space Museum in Alamogordo, New Mexico is well positioned to capture space tourism business in the years to come.

Not all space tourism involves short, near-Earth orbit flights like the Branson model. Some countries and organisations are working on ideas for orbital flights and longer duration travel. For example, Japanese engineers are developing a concept for a space elevator to take visitors to a space station, while there are plans mooted for space hotels (Laing & Crouch, 2004; Strickland, 2012). The Russian-designed Commercial Space Station could:

> host seven guests at a time in something less than five-star luxury:
> Dehydrated food will be served, and the bathroom is little more than a

suction tube. But the hotel will have unbeatable views of Earth from 250 miles up as it orbits at an average speed of 18,600 mph. After three months of training, guests will fly to the hotel on a Russian spacecraft and will be able to stay for as little as three days, or as long as six months. Russian authorities say the hotel could welcome its first guests as early as 2016. (The Week, 2012)

Space hotels in films are sometimes depicted as luxurious, with *The Fifth Element* (1997), for example, featuring a glamorous orbiting property above the mythical planet Fhloston, which is described as a 'hotel of a thousand and one follies', with twelve swimming pools. Guests watch a singer perform in a grand opera house. This is hardly representative of the type of facility that is likely to be built in space, given the realities and cost of constructing hotels in space. Strickland (2012: 903) argues that 'space tourists will have to be educated in what to expect'. Living spaces will be small and cramped, furnishings are likely to be spartan and functional, and food will be rehydrated and potentially lacking in flavour, due to the way the taste buds are affected in space. Going to the toilet may involve wearing a type of nappy on short flights, while the space loo won't be pretty (Figure 11.7). This will be far from the five-star experience many of these wealthy tourists will be used to on Earth. Explorer travellers might, however, find this lack of artifice to their liking, particularly those used to living out of tents or in the wild during an expedition.

Space tourists planning a stay in a space hotel may have to have medical screening before they are able to take up residence (Laing & Crouch, 2004), as well as training in the likely physiological effects of space flight, including space sickness (Strickland, 2012) and headaches due to the flow of blood away from the feet and towards the head. Longer vacations promise even greater health risks. These include loss of bone mass, greater risk of kidney-stone formation, anaemia and reduced cardiac mass and muscle atrophy (Laing & Crouch, 2004; Murphy, 2010). The dangers of exposure to high levels of radiation from cosmic rays are also very real, not to mention the risk of exposure to a solar flare. A recent study suggests that on a flight to Mars, radiation exposure would be such that it would be 'like receiving a whole-body CT scan every five or six days' (Tate, 2013).

NASA has been exploring ways to minimise these potentially harmful effects, including: regular exercise on a treadmill or exercise bike to strengthen bones and avoid problems with the cardiovascular system; the development of medication to minimise or eradicate these physiological changes; shielding individuals from radiation using water (which will be stored on board anyway or can be created); and creating new propulsion

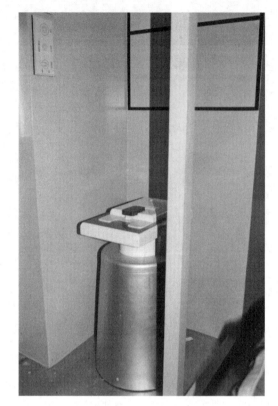

Figure 11.7 Toilet on the International Space Station
Source: J. Laing.

systems that reduce the time spent in space. These risks are a potential showstopper for a mission to Mars, where astronauts could be en route for three to six months, and need to be able to start work immediately when on the surface, without taking time out to recover from the effect of the space flight. Creation of artificial gravity on board (and in a space hotel) could avoid some of the problems associated with zero gravity. While this might destroy the fun of living in space, for those staying on the latter, 'a solution might be to have a special area set aside for experiencing weightlessness' (Laing & Crouch, 2004: 19).

New technologies may make space hotels more affordable, but not a great deal more comfortable. Bigelow Aerospace is building inflatable modules on the ISS, and is planning to create a space station from two modules – 'the world's first private space station' (Chang, 2013: 11). The cost reduction stems from the fact that it can be launched flat and then inflated in space. NASA's research on the ISS will examine how the modules stand up to space debris such as micro-meteorites (Chang, 2013). It may subsequently be used to create bases on the surfaces of the Moon and Mars.

While there are signs that space tourism has a market (Crouch & Laing, 2004), the idea that it automatically attracts risk-takers needs to be re-thought. Many explorer travellers we interviewed say they are put off the idea of space tourism by the amount of technology involved and thus the lack of control they would have over the experience. As Michael observes: 'Unless I'm going to start trying to put together a conglomerate who want to build a spaceship (laughs) to take you up there, you are going to have to go in someone else's [vehicle]. It would be a great journey but it wouldn't be the same adventure'. Keith similarly notes:

> I wouldn't want to go to space because it'll be so much in the control of everyone else. And if you go to space these days, I wouldn't be piloting it. I mean maybe one day, if I could buy my own spacecraft I'd have a go, but to me to be going there basically as a passenger, it would not give me the thrill. I would be very nervous [because] I wouldn't like having some-one else in control of my life. It worries me enough when I fly by an airline or go in a car with someone else. I like to be responsible for my own life and responsible for the risk-reduction that I do.

Harry says he would miss the hands-on, personal involvement he has with his other travel experiences on Earth:

> I think it's too removed from the senses. I mean I'd leap at the chance, if you were offering me a rocket to go to Mars, I mean wow. Yeah, it would give me enormous excitement but I can see that it's really within the realm of specialists and furthermore, a specialist who is guided by hun-dreds of other specialists with TV monitors and computers and so on.

While the Virgin Galactic plans deservedly hog the headlines and make space travel more affordable, these are still arguably high-end experiences. The parameters of space tourism are broader than that. They might encompass more affordable options such as zero-g flights, where the passenger on a para-bolic flight experiences a few seconds of weightlessness, and a package can be bought for US$4950 (Zero G Corporation, 2013), virtual space flight on simulators and using webcams and visits to terrestrial sites like space museums, theme parks incorporating space motifs, radio telescopes (Figure 11.8) and planetariums (Crouch et al., 2009; Laing & Crouch, 2004). These

Figure 11.8 Very Large Array in New Mexico – the telescopes used in the film *Contact*
Source: J. Laing.

might attract different market segments, although probably not the explorer traveller, due to the lack of autonomy and challenge.

Exploring Mars: Fiction Versus Reality

Living on a Mars base, on the other hand, might appeal to explorer travellers, due to the ability to explore the surface of the Red Planet. Being the first person to climb Olympus Mons, the highest mountain in the solar system, to see the Martian poles, or to visit the two Martian moons, Phobos and Deimos, would involve the requisite amount of challenge and bestow prestige on these individuals. Some of the concerns voiced about a space tourist's lack of freedom and control over their destiny would, however, apply in this context. Even though not strapped into a spacecraft, astronauts on Mars would be required to wear spacesuits when outside the base, for protection against hazards like dust as well as for life support. There is negligible atmospheric pressure on Mars and the air is about 96% carbon dioxide and thus toxic to humans (Murphy, 2010). They would also need to be contactable by Mission Control at all times. The level of independence would be very low, while the amount of risk avoidance required by those monitoring the mission would be high. This might frustrate the individual who revels in living on the edge (Lyng, 1990) and making their own decisions.

As discussed previously, the explorer traveller generally prefers a visceral experience, where authenticity is seen to result from using the minimum amount of technology possible. Rod spoke of these experiences as the most *pure* and was dismissive of another Antarctic expedition where the crew 'were in tractors with heaters and all that kind of stuff. To me, the purist way of being in the outdoors and not being cocooned from it and living it and feeling it is the primary thing; the primary driver'.

There will also be psychological issues relevant to sending people to Mars, for long periods of time and potentially permanently in some cases. It has been argued that the latter occurred when people were transported to Australia in the 19th century, with many unlikely to see their homeland and loved ones again. Yet they were still on the same planet and the risks were different. We really have no historical equivalent for someone heading off to live on another planetary body, although it is a popular subject for science fiction (e.g. *The Forever War*, *Lost in Space*). The impact of leaving Earth forever will no doubt require psychological counselling and careful selection of those who are best able to thrive in this kind of challenge. Kim Stanley Robinson's *Red Mars* (1993) depicts a psychologist who starts obsessively watching films about Earth, finding it difficult to cope with the idea of never seeing those places again. An individual on Mars, like those stationed in Antarctica, must deal with the sensory deprivation of an absence of greenery. They will never feel the wind in their hair or on their cheek. Living in close proximity with others will require an ability to overcome lack of privacy and deal with different personalities, again just like those living on an Antarctic base (Wood *et al.*, 2000).

On the bright side, unlike many science fiction versions of space travel, alien encounters will be limited on Mars to a likely discovery of a microbial presence and computers or robots may not run amok or take control (*Prometheus*, 2012; *2001: A Space Odyssey*; *Moon*, 2009). For those with an explorer heart, the opportunity to discover a largely uncharted planet may prove irresistible, in spite of the differences with respect to an earthbound expedition and the onerous requirements to toe the regulatory and administrative line. Once travel to Mars enters the private domain, rather than being sponsored or staged by government space agencies, the restrictions may relax to a degree, making them more attractive to explorer travellers. Nevertheless, some constraints will continue, such as the need to wear a spacesuit, at least until such time as the planet is terraformed. This is a controversial subject, making Mars essentially Earth-like, but one which Zubrin (1996: 218) argues 'is fundamentally a corollary to the economic viability of the Martian colonization effort'. It is also likely to happen naturally over time, as the human presence changes the environment, little by little. This is what we have done on Earth with climate change.

Why Explore Space?

Establishing a Mars base, along with a Moon base, would lay the groundwork for the human settlement of space. Arguments advanced for this step include the potential for greater understanding of the Earth and solving its environmental problems, and providing a contingency against the extinction of the human race from catastrophes like asteroid impacts or climate change (Cockell, 2007). There is also the rationale for space exploration which underpins much of the discussion in this book – the human fascination with the unknown and seeking new frontiers (Zubrin, 1996). Ultimately, the rationale for exploring space is really no different from that used to justify 19th century exploration, although the costs are substantially greater.

Leo, one of our interviewees, likes to see himself as one of the people expanding the boundaries of exploration: 'Someone's got to push. I mean the Wright Brothers obviously started something when they went flying. Christopher Columbus coming over here obviously started something. Someone's got to start this [space tourism] too, because you can see the results of it. This is just another step'. Space traveller Mark Shuttleworth agrees with these comments. He writes of his space tourism experience on the ISS:

> Space technology and the exploration of space have captured my imagination as long as I can remember. We are only just scratching the surface of the universe out there, and while we have plenty of work to do to get our affairs on earth in order, we can't lose sight of the need to keep exploring the whole of the universe in which we live. (Shuttleworth, 2002d)

It would be extraordinary if the urge to explore stopped at the Earth's surface. It would go against the historical trend for seeking the unknown and satisfying our curiosity for knowledge. If current research about the explorer gene is correct, we will not stop until we are headed for the stars. But to make this politically palatable, we may have to provide evidence of other social benefits, including technological developments that will help those back on Earth live more sustainably (Cockell, 2007).

12 The Explorer Traveller: The Myth Continues

In recent years, the tourism industry has wholeheartedly embraced the concept of the *Experience Economy*. As expounded by Pine and Gilmore (1999), this theory promises increased profits while improving the quality of the tourism product offered to tourists. The key is to move away from an old-fashioned view of services – accommodation, transport, tours – towards the planned creation of satisfying experiences. As this new paradigm has taken hold, there has been a clear shift in tourism marketing, with the promise of experiences – even profound and life-changing experiences – now ubiquitous.

Pine and Gilmore's theory is simply built on five propositions:

(1) As the service economy replaced a manufacturing-based one, so now the experience economy represents the next step beyond services.
(2) More complex and satisfying than mere services, experiences can be sold at a higher price, thereby increasing yield.
(3) There are four *experience realms*: entertainment, education, escapism and aesthetics. Ideally, *memorable experiences* occur in the *sweet spot* where all four realms intersect.
(4) Experiences are produced and staged like a play. The consumers are the audience who are entranced by the performance and actors. Surprising elements and special effects are included, 'staging the unexpected' (Pine & Gilmore, 1999: 96), to create the impression that this is a bespoke production.
(5) Experiences are scripted to control quality. Like a play, the actors read their lines while maintaining an illusion of naturalism.

Despite its widespread adoption, the experience economy is subject to criticism (see Chapter 10; also Frost & Laing, 2011; O'Dell, 2005; Ooi, 2005). It is essentially a supply-side view of tourism with a focus on how operators can increase yields through careful planning and staging. The emphasis on scripting seems at odds with what modern tourists want. Research into existential authenticity indicates that many tourists are searching for meaningful encounters and experiences that will make them feel more complete, more real (Cohen, 2004; Wang, 1999). Will a meticulously scripted and staged performance satisfy such consumers? Or is this the sort of production they will associate with Disneyland and Las Vegas?

We propose that many modern tourists want to go well beyond the experience economy. Yes, they want personally engaging experiences, but they do not want them constructed and scripted. Savvy modern tourists want engagements that they seek out, that they choose and even construct themselves. Nothing is a bigger turn-off for them than the idea that they are consuming a staged product. What they see as personally fulfilling is to *explore*.

There is a large segment of the modern tourism market that we describe as the *explorer traveller*. This grouping is defined by the following characteristics:

(1) This is a broad group, which is growing. Educated, urban and with reasonable incomes, they value regular travel as a key component of their lifestyle and identity. They are not just backpackers or drifters, but cover a wider range.

(2) They are *fully independent travellers*, who reject tours, mass tourism and the manufacturing and scripting of experiences.

(3) They draw inspiration from both historical narratives of explorers and accounts by modern-day travellers. They are great consumers of media, ranging from books and cinema through to social media. They rely on and trust *organic* media sources and, in turn, are sceptical of advertising copy churned out by destinations and operators.

(4) They link travel to discovery – which they see in two ways. The first is visiting new places and gaining an understanding of them. If a place is already very popular, they either are not interested, or they want to go beyond the surface and find out something different about it. There are overlaps with ecotourism and cultural tourism in this search for knowledge and meaning.

(5) The second sense of discovery is intensely personal. The explorer traveller sees their journey as an achievement, possibly a test and potentially transformative. In seeking out new places and experiences, they hope to effect personal change and growth. The concept of the search for existential authenticity is very important for explorer travellers. They want *real* experiences that will change them for the better. Such a personal

path can only be achieved as an individual or small group, reinforcing notions that tours and scripted experiences are to be avoided.

(6) Mediation is a difficult concept within this quest for personal discovery. The explorer traveller delights in doing it their own way, from researching the trip to stumbling across unknown attractions and eating places. The normal mass tourism guide (and their pre-packaged interpretation) is not satisfactory. Mediation is only acceptable from local sources, preferably non-professional with few ties to the mainstream tourist industry. Meeting with locals and sharing meals and everyday life are attractive – albeit if only for a few days.

The Demonstration Effect

The ranks of elite explorers are quite small. In contrast, their impact is large. They produce a *demonstration effect*. Media focusing on their adventures, heroics and self-actualisation are widespread. In the 19th century, their stories were consumed through books, magazines and newspapers. The 20th century saw a broadening of access through cinema, radio and television and in our current times the internet and social media have disseminated tales of elite explorers. These seductive narratives filter down to everyday people and have popularly become something that readers and audiences are familiar with and associate with adventure. Most importantly, the myth of the explorer has become so prominent in modern society that many people want to duplicate it. Having had it demonstrated through the media, they too wish to have the same experience.

The demonstration effect is apparent in the accounts of modern explorers. It can be seen in our interview with Graham, the Outback trekker. When we asked him whether anyone had inspired him, he responded, 'no, just books I've read over the years probably ... but I can't point to any individual who inspired me'. However, as our interview proceeded, he opened up, talking about Edmund Hillary, James Cook, the Mt Everest tragedy (about which he had read six books) and Augustus Gregory (whose journey he recreated). One explorer's account stood out for him:

One of the first books I can ever remember reading was Thor Heyerdahl's *Kon-Tiki* excursion and it was enormously influential. I must have read it about ten times. About how exciting it was to do these types of things and Thor Heyerdahl was an interesting fellow because it's not like reading about someone who lived four or five hundred years ago. He was still alive ... and doing these wonderful things.

And so that left a deep impression on me and ... he was doing something scientific. He had a theory about the way Polynesians populated the Pacific and he was an anthropologist, so instead of sitting in a library ... he ... actually built a big balsa raft and sailed it across the Pacific. And you read his books, all of them, they're a combination of adventure story and scientific things, and so that's what I really like and I found very inspiring about what he did. And so I suppose I started out trying to do the same thing.

In turn, the demonstration effect continues through Graham writing books and magazine articles and regularly appearing on a radio programme. Those who hear or read him are potentially inspired to explore. They might want to recreate his experiences, or do something similarly adventurous in a different place, or undertake travel that they find personally challenging. The last of these might not be at anywhere near the level of Graham's expeditions, but nevertheless might be difficult and adventurous for the person inspired to undertake it. It is through this last step that there might be a significant magnification of the impetus to explore. The account of one elite explorer might encourage hundreds, even thousands, to engage in their own explorer travel.

While our focus is on written narratives by explorers influencing others to follow in their footsteps (metaphorically, if not literally), there are other media at work. Cinema, for example, is worth considering in how it presents the explorer myths. During the course of writing this book, we viewed a number of films that represented explorer travel as desirable, testing and worthy. Two films made in 2012 – *Prometheus* (science fiction) and *The Hobbit* (fantasy) – both commence with the uncovering of a lost map, sparking dangerous expeditions into the unknown. *The Life of Pi* (2012) follows the novel (discussed in Chapter 7) in detailing the adventures of a boy cast adrift after his boat sinks. *Gravity* (2013) similarly features an explorer traveller cast adrift in a highly hostile environment. In this case, after the destruction of their spaceship, stranded astronauts have only hours to return to Earth. *Kon-Tiki* (2012) reconstructs the real-life voyage of Thor Heyerdahl, with particular emphasis on the hero's motivation. Similarly, *Tracks* (2014) brings the account of a real explorer traveller (Robyn Davidson) to the screen. Finally, we found *Midnight in Paris* (2011) a particularly intriguing example of a hero's journey and trials, albeit in an urban context.

The hero of this Woody Allen film is Gil (Owen Wilson). Well off, educated and urbane, he seemingly has everything. He takes a trip to Paris with his girlfriend Inez (Rachel McAdams) and her parents. However, Gil is troubled. He is successful as a screenwriter, but now finds that unrewarding. He would like to be a novelist. Part of the motivation for the trip is to allow him to make a decision. Inez is unsupportive; she wants him to stay in his comfortable well-paid job. Echoing her materialism, they stay in an expensive international hotel that could be anywhere in the world, a cross between the *Accidental Tourist* and *Sex in the City*. Gil and Inez are juxtaposed as different types of travellers. Happy at home, travel brings out how different they are. Inez is very much the mainstream tourist, set up to be scorned by the audience.

Disillusioned, Gil takes to wandering his beloved Paris at night. He is a *flaneur*, drawn to the haunts of his literary heroes – the *Lost Generation* of the 1920s. Here again we see both the demonstration effect and intertextuality at work. He is drawn to places that he has not only read about, but where the writers he admires once lived and worked. Magically, as he walks each night he is transported back to that time, mixing with the likes of Ernest Hemingway and Gertrude Stein. During the day, he also wanders throughout Paris. On the banks of the Seine he meets Gabrielle and they find they share a common love of the city's history and literature. Through these peregrinations, Gil gradually works out the issues in his life. He will move to Paris and become a novelist and he will leave Inez and take up with Gabrielle.

The Inner Journey

Many explorer narratives emphasise the inner journey over the ostensible goal – the testing of character and perseverance, the need to look inwards and understand oneself better. When we interviewed Michael, he reflected upon this and how he saw his journeys:

I think it's always been about self-exploration, it's always been about the adventure of not knowing whether you're going to be able to achieve what you set out to achieve, what's within yourself rather than the attainment of some kind of goal. Sure, you want to get to the peak, the top of the peak – but with all due respect – the summit of Everest has been photographed many, many times, as has the view. When Peary went trying to beat Cook to the North Pole, they knew it was a floating piece of ice on the ocean at 90 degrees north, they knew they weren't going to discover a new culture . . . they knew it was just frozen ocean. But they still forced themselves on. So I don't think *exploration* has anything to do with it. I think it's *self-exploration*. That's why I do it. I keep finding myself going back looking for the part of my own personality that comes out when I do these trips. It's a more resourceful side of me that comes to the fore when I do these things ... you have to go back and put yourself in risky situations or challenging situations to find that person.

Such sentiments suggest a new paradigm for exploration, replacing conquest with finding balance and satisfaction. This quest for personal development, we argue, is a defining characteristic of explorer travel. Furthermore, it is present in a wide range of types of experience. In line with Cohen (2004), we see a continuum of explorer travel experiences, ranging from the elite to the everyday. Whether a travel experience qualifies as exploration is a sub-jective judgement, one in the eye of the individual traveller. Michael (and this was echoed by others we interviewed) explained this subjectivity in the following way:

I think the spirit of human endeavour and the spirit of adventure man-ifest themselves in different ways in different people ... anything where you feel as though you're stepping outside of your comfort zone represents an adventure for you, whatever that happens to be. And only you will know whether something is an adventure for *you* or not. One person's adventure may not even cause another person's heart to skip a beat ... walking along a boardwalk to a mountain in Tasmania ... for some people that is an adventure. They're not sure if they're going to be able to get to the end of it. Nothing to do with the fact that people have been there before them, they just don't know whether they're going to be able to achieve it and that's the adventure for them.

By emphasising inner discovery and development, a wide range of travel experiences may be counted as explorer travel. At one end of the continuum are a small number of elite adventurers. In contrast, we conceptualise that the opposite end of the continuum contains larger numbers of ordinary people undertaking travel that challenges them. To illustrate this, we con-sider two examples of explorer travel: walkers on the Camino Santiago and *Food Explorers*. Both are increasing in popularity and appeal to travellers who are seeking personal change.

Walking the Camino Santiago

In 2013 we visited the Spanish pilgrimage town of Santiago de Compostela. It was not the culmination of a pilgrimage walk (on this occasion!) but was merely a bus trip at the end of a conference. We had a long-standing interest in the Camino Way, the pilgrimage that ends up at Santiago Cathedral, the resting place of St James. One of the Apostles, he had proselytised to Galicia, with limited success, in the period after Christ's crucifixion, before returning to Jerusalem in AD 42 and martyrdom at the hands of King Herod. His body was allegedly returned to Spain and buried in Galicia. According to the legend, in the 9th century lights were seen, and a tomb was discovered, said to be that of St James. A cathedral was built, and the region became a focus for Christian pilgrimage for centuries. By the 19th century, this had dwindled to a few dozen pilgrims a year (Xacobeo, 2010). It has had a resurgence of interest in the 20th century (González & Medina, 2003), perhaps linked to a decline in organised religion in the West, which created a hunger for spiritual experiences and a desire to find a deeper meaning or purpose in one's life. Between 1986 and 2005, on average 100,000 pilgrims received a Compostela Certificate each year, given to those who show evidence of having walked at least 100 kilometres or ridden 200 kilometres, through stamps in one's *Credencial del Peregrino* or Pilgrim's Passport (Wells & Wells, 2008). The number of pilgrims spike in Holy Years, when Apostles Day (25 July) falls on a Sunday. Traditionally, in Holy Years, pilgrims receive a full indulgence for their sins and the Great Door or *Puerta Sancta* is opened in the Cathedral. The last Holy Year was 2010, when 272,135 people received a Certificate.

We felt the excitement of the others around us in Santiago Cathedral, as they crowded in with their walking gear, staffs and scallop shells, the symbol of St James, after completing their pilgrimage. There are eight Ways of St James (Xacobeo, 2010). Some had walked the French Way from Saint Jean-Pied-de-Port, the so-called classic route, which is nearly 800 kilometres and early on traverses the Col de Lepoeder at 1450 metres, before entering Spain. This Pyrenees crossing is serious stuff – a French woman died of exposure here in 2009, and Wells and Wells (2008) describe how a fellow walker has her contact lens ripped out of her eye by the wind. The Ways follow the various routes that pilgrims took to connect with the Iberian Peninsula (Xacobeo, 2010), including the English Way and the Portuguese Way. We stumbled onto the Portuguese Way in Porto, with arrows (Figure 12.1) headed one way to Santiago and in the other direction towards Fatima, another famous pilgrimage site. These arrows, along with markers and scallop shells, help the pilgrims stay on the correct route. We followed them for a way, and fantasised about what it would really be like to be on a journey to Santiago. Wells and Wells (2008: 129) describe the arrows, painted on everything from fences to curbs as 'reassuring and welcoming'. The occasional unintentional detour occurs, but this is part of the challenge.

Figure 12.1 Arrows in Porto showing Santiago and Fatima, the great pilgrimage centres
Source: J. Laing.

Several things occurred to us as we strolled around the streets of Santiago and witnessed the great *Botafumeiro*, a swinging incense burner pulled by ropes, which arcs and loops its way through the nave of the Cathedral. The first was that Jennifer felt a bit like a fraud. She had expected the visit to the Cathedral to be highly emotional and strongly significant. Instead, she felt flat. Not having done the pilgrimage walk, the experience was not *authentic*. She felt that we didn't deserve to be here. The road of trials was an integral part of the experience, not an add-on. This echoed some of the comments of the people who had written about their experiences on the Camino Way. Best and Bowles (2007) are scathing about people who join the pilgrimage close to Santiago and have their luggage sent on by a driver, or are even driven or take the bus between stages, while still collecting their stamps in their Passport. Eli rages: 'Is this what Led Zeppelin meant about buying a stairway to heaven? Everybody is racing to have their sins forgiven, and they don't care how many they have to commit in the process' (Best & Bowles, 2007: 268). Colin Bowles hates the crowds racing to the next *albergue* or hostel and taking short cuts. He talks to a Spaniard who:

> has travelled all the way from Roncesvalles as we have, and is now turn-ing back in disgust. Some may call him a purist but he has a point. In Castañeda they told us that the twelfth-century lime ovens were once

used to make the cement for Santiago's cathedral, and medieval pilgrims each carried a heavy stone all the way from Triacastela for use in these kilns. These days they don't even carry their own packs. (2007: 280)

It seems there is a *right* way to do this. Getting the Compostela without doing the hard yards is *cheating*. Even the use of the pilgrims' hostels or *albergues* is controversial. Hape Kerkeling (2006) is a German comedian who walks the Camino Way, and after his first night in a hostel, vows never to repeat it. He cannot sleep with the snoring and recognises that he will not complete his walk if he doesn't get a good night's sleep. Kerkeling observes that many of the people sleeping in the *albergues* could well afford a more luxurious standard of accommodation. They are penny-pinchers, spoiling it for others, who may find it difficult to find a hostel bed in some places. This echoes a comment made to us by our guide in Santiago, who felt that some people walk the Camino Way for a cheap holiday, and don't travel in the spirit of a true pilgrim. In fact, Compostelas are only awarded to those who can state that they have 'some spiritual purpose. If not, your Compostela is demoted to a measly *certificado*, and there's no getting into heaven with one of those' (Best & Bowles, 2007: 26). In one sense, the idea of a Compostela, with authorities regulating the walk with a passport and stamps, and individuals being required to substantiate how far they have walked and what their motivation was, is similar to the phenomenon of explorer travellers having their exploits verified by an independent body. Brown (2001: 51) refers to the 'quaint ethics of polar adventure' with respect to where the journey should commence. They 'ordain that starting on the real edge of the continent, where the permanent ice meets the open sea in high summer, is pure and superior, because this is the closest the explorers of the heroic age could get without aircraft. Thus Berkner Island is favoured by those who wish their traverse to the Pole to remain untainted and can afford it'. Some explorer travellers, however, feel that this intrudes on their freedom to choose their own route. Rod, one of our interviewees, found this overly regulated attitude in the context of polar trekking disappointing:

[They are] imposing on the adventuring world, the Polar adventuring world, a template of how things should be done ... And if they don't follow that template, they're using terms like 'disqualified', they're using terms like 'not a proper expedition'; all this kind of stuff which leads to dangerous things ... It doesn't really matter if you actually reach the Pole or not or whether you start from this point. If you have a Website and you tell people about it, just say where you started from and say where you finished, if that's important to you.

Figure 12.2 Pilgrims, wearing the latest hiking gear, queue for the tomb of St James in Santiago Cathedral
Source: W. Frost.

The second observation we made was how clothing seemed to define these pilgrims. They were all dressed in the latest hiking gear and labels (Figure 12.2). It became a marker of identity, along with the staffs and scallop shells. The shells are given to each pilgrim when they receive their Passport. Interestingly, Wells and Wells (2008: 29) originally felt awkward displaying their shell to others on the route: 'There can be a reserve, a sort of shyness in openly declaring that you are on a pilgrimage.' Eventually, however, 'the shell became part of us and engendered an unspoken bond with other walkers'. There is a reference here to the *communitas* which often grows between explorer travellers who undergo the same experiences, at least the same *authentic* ones. The symbol becomes something that is earned and displayed with pride and denotes a real explorer traveller, someone who has undergone a visceral, difficult and transformative experience and achieved something momentous. Interestingly, the shell and staff have also been commodified through souvenirs such as T-shirts and fridge magnets, many featuring pop culture icons such as Hello Kitty, Homer Simpson and SpongeBob SquarePants dressed in the traditional pilgrim costume (Figure 12.3). Such commodification is despite the fact that 'many pilgrims understand themselves to be rejecting consumerist lifestyles, even if only temporarily, while on the Camino' (Norman, 2009: 68).

Figure 12.3 Hello Kitty pilgrim T-shirt sold in the souvenir shops of Santiago de Compostela
Source: W. Frost.

The sense of spirituality that pilgrims find on the route gives the experience meaning and purpose. Even suffering is purposeful, as it leads to greater awareness of boundaries, self-imposed or otherwise, and humility. No one is invincible, and the traveller has to overcome doubts and fear to keep going. Kerkeling (2006) likens his travel to a prayer. He was initially unsure whether the walk would offer any sort of spiritual enlightenment but, as it unfolded, he found time for contemplation and the opportunity to put his life, including the disappointments and grievances, into perspective.

For Eli Best, it is 'a leap of faith that invokes a sensory overload so intense that it's only after the event – when your feet are healed or safely on the ground – that you can look back and realise what actually happened. Suddenly, why you did it in the first place isn't as important as the experience itself' (Best & Bowles, 2007: 185). Her walking partner Colin Bowles, through his walk, starts to come to terms with tragedy in his life, although he recognises that the experience did not give him 'a last minute miracle' (Best & Bowles, 2007: 313) or a quick fix. It is just one step in a long journey: 'Out here I have felt, if only for a time, small and humble, and have heard in ancient stones the prayers of pilgrims long since dead. I have felt – if for a moment – a part of eternity, and a greater plan, at the end of the Earth' (Best & Bowles, 2007: 314). Humility is an element of an explorer travel experience, particularly when faced with the grandeur and power of nature. Eric Weihenmayer (2001: 283) likes this aspect of climbing a mountain: 'It is a realm where humans haven't reached godlike status, a realm that demands humility. Human frailty is amplified, human ambition nullified'.

Wells and Wells emphasise the inner journey they made on the Camino Way, and the relationships with others. This experience was both fulfilling and 'liberating'. Like Bowles, they acknowledge the part the walk will play in 'our continuing journey' (Wells & Wells, 2008: 162). It has *enriched* their lives immeasurably. This also reflects Aaron's view of the ongoing influence of his solo North Pole trek on his life: 'It's of course [a] profound experience, and I always believed at the time that it will be an experience that I'll continue to live on for the rest of my life. It's not just relegated to that moment in time'.

While not all pilgrims might be explorer travellers, there are clear crossovers between them and some might be characterised as adventurous pilgrims. They both approach their journeys as potentially life-changing and immerse themselves in the experience, deriving a strong sense of satisfaction and fulfilment from their achievements, both mental and physical. These are journeys that touch the soul.

Food Explorers

As an outcome of our research on regional food and wine, we have introduced the idea of the *Food Explorer* (Laing & Frost, 2015, forthcoming). This is a niche tourism trend which is increasingly becoming important for certain destinations. Food Explorers – like explorer travellers in general – tend to be urban, educated and with high levels of income. They are independent travellers who are keen to experience food and wine that is distinctive of a place and they enjoy discovering and learning about different food styles, traditions and heritage (Figure 12.4). Engaging with food culture is a major determinant in their holiday and destination choices. As Quan and Wang (2004) argue, in the past, food has tended to be dismissed as merely a *supporting* experience in tourism, whereas there is now a modern trend towards it being a *key* experience.

We identify three main hallmarks of the food explorer and note that these could be easily adapted to other forms of explorer travel:

(1) Food explorers do not appreciate a commodified tourist experience. They eschew guided tours and tourist menus, preferring to eat like locals and discover for themselves what the destination has to offer.

Figure 12.4 Discovering a new restaurant and bar in Salamanca Place, Hobart
Source: J. Laing.

They want surprises rather than certainty. For example, the food explorer is delighted by the prospect of eating at a local restaurant where many menu items are unfamiliar and they may not know much of the language. This is not only fun but is valued for its strong educative value.

(2)   They place great value on experiencing food and drinks that are authentic to a place. They seek out local produce and regional specialities. These are particularly prized if they are little known or distinctively limited to a particular area. A key part of their exploration is wandering or *flaneuring*, chancing upon markets, seasonal produce and restaurants, cafes and bars staffed and patronised by locals.

(3)   They are deeply concerned with cultural and environmental sustainability (and here a parallel may be drawn with ecotourists). They are knowledgeable about issues and debates relating to food and culture and they want to put their knowledge into action. Ideas and trends of strong interest to them include organic food, localism, slow food (Figure 12.5) and maintaining food traditions.

Certainly food explorers may be seen at the *soft* end of the explorer traveller continuum, for their experiences offer little risk. However, we argue they represent an important adaptation of the explorer myth. They are a growing trend of travellers who deliberately reject mass tourism and the experience

Figure 12.5 Promoting slow food in a restaurant in Pistoia, Italy
Source: J. Laing.

economy and seek to fashion their own experiences, stepping outside their comfort zone and entering the unknown.

A Research Agenda for Explorer Travel

Through our study, we have identified 10 key areas which we argue might provide a starting point for future research.

(1)   Does the Experience Economy model need to be rethought in light of our findings with respect to the explorer traveller? We argue that they crave authentic rather than staged experiences, prefer to travel independently, and will pay a premium for these extraordinary experiences provided they are able to play a role in co-creating them. This does not fit neatly within any of the dimensions of the model as it stands. It may need to be expanded to include a new dimension – exploration – or alternatively

updated to re-conceptualise tourist experiences as *organic and dynamic*, rather than heavily scripted and pre-packaged phenomena.

(2)   We argue that explorer travel occurs along a continuum and provide examples in this book of travel to adventurous places on the frontiers of our world, tourists walking along the Camino Way in Spain and food explorers. Future research could examine examples of explorer travellers in other contexts, such as cultural tourism, heritage tourism or dark tourism experiences.

(3)   What role does spirituality play in these explorer travel experiences? Swarbrooke *et al.* (2003: 9) outline the 'core characteristics' or essential qualities of adventure but do not include spirituality. This might be true in some adventure tourism contexts such as bungee jumping, skydiving or white-water rafting, yet our work suggests that there is a strong spiritual element in explorer travel experiences, particularly in desert, mountain or polar environments, perhaps linked to romantic views of the sublime. Space tourism might also be a potential transcendent experience, based on some of the comments of our interviewees. Future studies could explore in more depth the spiritual nature of these types of journeys, which might resemble a pilgrimage or sacred journey for some individuals.

(4)   We would have liked to have interviewed more women in this study, exploring their reasons for travelling to the peripheries, in what has been for a long time a male-dominated activity. Several women we interviewed noted the male bias and discussed reasons why this might be so. Future studies could employ feminist methodologies and attempt to give the female explorer traveller a greater voice, as well as examining the potential of this kind of experience for transformation and emancipation.

(5)   Many explorer travel experiences considered in this book involve the individual facing sustained periods of extreme, highly dangerous and testing activities. What is the attraction of these hellish experiences, how might the mythic construct of the *katabasis* assist us in understanding them, and does this desire to experience deep suffering have spiritual undertones?

(6)   What role does culture play in these experiences? The frontier travellers taking part in this study were essentially English-speaking and of an Anglo-Saxon, white background. It might be useful to consider whether certain races or cultures tend to embrace this kind of travel experience more than others and whether this is linked to history or heritage, particularly a colonial history and shared tropes within a cultural boundary, and/or the access to funding or opportunity. Cross-

cultural studies of explorer travellers might be valuable, given the cultural limitations of our study. Issues of translation and mistranslation are also worthy of further research.

(7) Space tourism is likely to burgeon over the next decade, once the spaceport in New Mexico is completed and Virgin Galactic flights begin. We still know very little about the potential space tourist, including their expectations of these experiences, whether this is influenced by books, television and films, and the possible disappointment which might occur due to the lack of freedom, control and active participation, at least in the early years of space tourism. There is also scope to examine whether there is widespread interest in travel to the Moon and Mars, should the technology be in place to allow this to happen, and what activities and attractions might interest a tourist when they arrive.

(8) Explorer travel experiences appear to exhibit the elements of Seligman's (2011) theoretical model of wellbeing known as PERMA (positive emotions, engagement, relationships, meaning and achievement). They are not mere thrill-seeking or hedonistic activities, but leave a profound effect on these individuals, and may also have benefits to health, quality of life and life satisfaction. Saunders *et al.* (2014) have found this model to be a useful lens for examining long-distance walking. Further research applying this model but also other theories of positive psychology to the explorer travel context may help us better understand the attraction of these experiences, but also their potential positive impacts on the lives of these travellers.

(9) Is there an explorer gene and how does it affect or influence the travel experience? Are explorer travellers more or less likely to carry this gene? Teasing out these questions would require genetic research, potentially on an international scale. An allied question would be to consider whether explorer travellers share similar personality traits and whether this too has a genetic base.

(10) Our study suggests that explorer travel can be understood using mythic constructs, such as Campbell's (1949) hero's journey and the Greek *katabasis*. A similar framework which might apply in this context is the *catharsis* (emotional release), which has its roots in Greek drama. Future research could examine how these journeys intersect with modern myth through narratives, particularly books, and potentially other forms of influential modern media such as cinema and documentaries. Re-enactments might be an interesting context for this research, given the role that myth apparently plays in these experiences.

The Cycle of Myth-Making

When we started writing this book, we saw it as a natural sequel to our first collaboration, *Books and Travel: Inspiration, Quests and Transformation* (2012). In that book, we looked at those individuals who seek out adventure, the explorer travellers, but also noted in many instances the critical role that books play in inspiring this process and shaping their imaginings of travel. We have highlighted the mythic dimension of many of these journeys, and used Campbell's (1949) hero's journey as a theoretical framework for understanding how these experiences are constructed and the meanings that are derived from them. The explorer traveller is the hero in their own story and must undergo and overcome hardship and trials in order to attain a boon or reward and return home, irrevocably changed by the experience. Our work thus supports Campbell's (1949) assertion that the hero's journey is a *monomyth* (Laing & Frost, 2012), ubiquitous across cultures and eras.

Explorer travellers in turn often write their own accounts of their experiences, creating an endless cycle of myth-making. New forms of technology might be used to disseminate these accounts, such as blogs, Twitter, Facebook and personal websites, but the power of the narrative remains the same. The *frontier* is one modern myth. Another is the *explorer myth*, a Western cultural legacy which remains vibrant in the 21st century. In a world where seemingly more of life is lived virtually than face-to-face and we are more litigious and risk-averse than ever before, seeking to coddle our children against any dangers or disappointments and retreating to our urban cocoons at night, it might be surprising that explorer travel is an emerging trend in tourism. Perhaps this is a reaction to our over-regulated and technology-driven lives, but might also reflect the deep need in some human beings, which might be genetically derived, to experience life in the raw, with all its dangers and delights. This trend is not just the case in an adventure context, as discussed earlier in this chapter, but appears to be a trend more broadly. A group of travellers are looking for authentic, sustainable and organic experiences, beyond the sheltered mass tourist bubble or staged experience, which are deeply immersive and provide sufficient challenge and opportunities for discovery and personal growth. In the future, they may be bound for interstellar space as the human desire to move beyond the known and the everyday continues apace.

Sources A: Participants Interviewed

Pseudonym	Country	Explorer travel experiences
Aaron	USA	North Pole trek; cycle across USA
Alex	Australia	Round the world solo sail
Andrew	Australia	Mountain climbing including Mt Everest
Brett	Australia	Cycle across Australia from south to north; Australian desert trek
Bryan	The Netherlands	Mountain climbing (mainly solo) including Mt Cook
Charlie	Australia	Australian desert treks
David	Australia	Mountain climbing including Mt Everest
Doug	Australia	Mountain climbing including Mt Everest
Emily	USA	Potential space tourist; works for X-Prize company
Evan	UK	Balloonist; potential space tourist
Geoff	UK	Row across Pacific Ocean; North Pole treks
Graham	Australia	Australian desert treks
Harry	UK	Desert treks; jungles; North Pole journey
Helen	Australia	Mountain climbing including Mt Everest
Jack	Germany	Round the world adventurers, including desert treks
Charlotte	Germany	Round the world adventurers, including desert treks
Janine	Belgium	Mountain climbing – Seven Summits including Mt Everest; desert treks; Antarctic trek
Jonathan	Australia	Antarctic, including South Pole treks
Karen	Australia	South Pole trek
Keith	Australia	Balloonist; mountain climber; round the world solo aviator
Leo	Canada	Potential space tourist; works for X-Prize company

Pseudonym	Country	Explorer travel experiences
Martin	UK	Mountain climbing – Seven Summits including Mt Everest
Max	Australia	Backpacking in Nepal
Mitchell	UK	North Pole treks
Melinda	Australia	South Pole trek
Michael	UK	North Pole trek; South Pole trek; desert treks; kayaking
Murray	UK	Mountain climbing including Mt Everest
Paul	Australia	Round the world solo sailing
Peter	Australia	Deep sea diving, including ocean wrecks; cave diving
Richard	Australia	Diving, including Polar diving; North Pole treks
Robert	Australia	Round the world solo sailing; Antarctica
Louise	Australia	Sailing; Antarctica
Rod	Australia	South Pole trek; North Pole treks
Ross	Australia	Sailing down Amazon and Yenisey rivers; kayaking
Sarah	Australia	South Pole trek
Sean	UK	Potential space tourist; director of X-Prize company
Simon	Australia	Deep sea diving, including ocean wrecks; cave diving
Wayne	USA	Travelling through the US West
Will	Australia	Mountain climbing including Mt Kilimanjaro

Sources B: Primary References

Explorer Travellers

Aldrin, B. with Warga, W. (1973) *Return to Earth*. New York: Random House.

Allen, B. (1992) *The Proving Grounds: A Journey into the Interior of New Guinea and Australia*. London: HarperCollins.

Amundsen, R. (1912) *Race to the South Pole*. Excerpts from *The South Pole: An Account of the Norwegian Antarctic Expedition in the 'Fram' 1910–1912* (trans. from Norwegian by A.G. Chater) (2007 edn). Vercelli: White Star.

Anker, C. and Roberts, D. (1999) *The Lost Explorer: Finding Mallory on Mount Everest*. New York: Touchstone.

Arneson, L. and Bancroft, A. with Dahle, C. (2003) *No Horizon Is So Far: Two Women and Their Extraordinary Journey Across Antarctica*. Cambridge, MA: Da Capo.

Asher, M. (1988) *Two Against the Sahara: On Camelback from Nouakchott to the Nile*. New York: William Morrow.

Best, E. and Bowles, C. (2007) *The Year We Seized the Day* (2010 edn). Sydney: Arena.

Blixen, K. (1937) *Out of Africa* (1986 reprint). London: Penguin.

Bird, I. (1879) *A Lady's Life in the Rocky Mountains* (1991 reprint). London: Virago.

Bohong, J. (1989) *In the Footsteps of Marco Polo*. Beijing: New World Press.

Boukreev, A. (2001) *Above the Clouds: The Diaries of a High-Altitude Mountaineer*. New York: St Martin's Press.

Bowermaster, J. (2000) *Birthplace of the Winds: Adventuring in Alaska's Islands of Fire and Ice*. Washington, DC: Adventure Press.

Breashears, D. (1999) *High Exposure: An Enduring Passion for Everest and Other Unforgiving Places*. Edinburgh: Canongate Books.

Brown, I. (1999) *Extreme South: Struggles and Triumph of the First Australian Team to the Pole*. Terrey Hills: Australian Geographic.

Burnaby, F. (1877) *A Ride to Khiva: Travels and Adventures in Central Asia* (1972 reprint). London: Charles Knight.

Burton, R.F. (1855) *Personal Narrative of a Pilgrimage to El-Medinah and Meccah* (2005 reprint). London: Folio.

Burton, R.F. (1872) *Zanzibar: City, Island and Coast (Vol. 1)* (1967 reprint). London: Tinsley.

Cherry-Garrard, A. (1922) *The Worst Journey in the World* (2010 reprint). London: Vintage Books.

Christie Mallowan, A. (1946) *Come Tell Me How You Live* (1975 reprint). London: Collins.

Cope, T. and Hatherly, C. (2003) *Off the Rails*. Camberwell: Penguin.

Cremony, J.C. (1868) *Life Among the Apaches* (1983 reprint). Lincoln, NE and London: University of Nebraska Press.

Dalrymple B.C. (1907) *In the Footsteps of Marco Polo: Being the Account of a Journey Overland From Simla to Pekin*. Edinburgh and London: William Blackwood.

Dalrymple, W. (1989) *In Xanadu: A Quest* (2012 edn). London: Vintage.

Danziger, N. (1988) *Danziger's Travels: Beyond Forbidden Frontiers*. London: Paladin.

Davidson, R. (1980) *Tracks*. London: Picador, Pan Macmillan.

Davidson, R. (1996) *Desert Places*. London: Viking.

Dickinson, M. (1998) *The Death Zone*. London: Arrow.

Eames, A. (2004) *The 8.55 to Baghdad*. London: Bantam.

Fermor, P.L. (1977) *A Time of Gifts*. London: Penguin.

Ferreras, P. (2004) *The Dive: A Story of Love and Obsession*. London: CollinsWillow.

Fiennes, R. (1993) *Mind Over Matter: The Epic Crossing of the Antarctic Continent*. London: Sinclair-Stevenson.

Goddard, J. (2001) *The Survivor: 24 Spine-Chilling Adventures on the Edge of Death*. Deerfield Beach, FL: Health Communications.

Grann, D. (2009) *The Lost City of Z: A Legendary British Explorer's Deadly Quest to Uncover the Secrets of the Amazon* (2010 edn). London: Simon & Schuster.

Grylls, B. (2000) *Facing Up: A Remarkable Journey to the Summit of Mount Everest*. London: Macmillan.

Hamilton, C. (2000) *South Pole 2000: Five Women in Search of an Adventure*. Sydney: HarperCollins Publishers.

Hare, J. (2002) Surviving the Sahara. *National Geographic* 202 (6), 54–77.

Harlin, J. (2007) *The Eiger Obsession: Facing the Mountain That Killed My Father* (2008 reprint). London: Hutchinson.

Harper, A. (1999) Stage 6 – heading for the sign. *Capricorn Expedition, The Journals*. See http://www.capricornexpedition.com/stage6.htm (accessed February 2005).

Hartley, C. with Y. Chang (2002) *To the Poles (Without a Beard): The Polar Adventures of a World Record-Breaking Woman*. London: Simon & Schuster.

Heller, P. (2004) *Hell or High Water: Surviving Tibet's Tsangpo River*. Crows Nest: Allen & Unwin.

Hillary, P. and Elder, J.E. (2003) *In the Ghost Country: A Lifetime Spent on the Edge*. Milsons Point: Random House Australia.

Hofmann, C. (1998) *The White Masai* (2007 edn). New York: Amistad.

Jarvis, T. (2013) *Sir Douglas Mawson's Antarctic Journey*. See http://www.timjarvis.org/expeditions/sir-douglas-mawson-antarctic-journey/ (accessed May 2013).

Jellie, D. (2012) Climb of reckoning. *The Age*, Travel section, 10 November, pp. 16–17.

Kelly, K. (2000) *Hard Country, Hard Men* (2002 edn). Alexandria: Hale & Ironmonger.

Kelly, K. (2003) *Tanami: On Foot Across Australia's Desert Heart*. Sydney: Pan Macmillan Australia.

Kerkeling, H. (2006) *I'm Off Then: Losing and Finding Myself on the Camino De Santiago* (2009 translation). London: Free Press.

Kingsley, M. (1897) *Travels in West Africa* (2007 reprint). London: Folio Society.

Kozel, B. (2002) *Three Men in a Raft: An Improbable Journey Down the Amazon*. Sydney: Pan Macmillan.

Krakauer, J. (1997) *Into Thin Air*. New York: Anchor.

Laing, O. (2011) *To the River* (2012 edn). Edinburgh: Canongate.

Langford, N.P. (1905) *The Discovery of Yellowstone Park: Journal of the Washburn Expedition to the Yellowstone and Firehole Rivers in the Year 1870* (1972 reprint). Lincoln, NE: University of Nebraska Press.

Lewis, D. (1975) *Ice Bird: The First Single-Handed Voyage to Antarctica*. Glasgow: Fontana.

Lindblade, A. (2001) *Expeditions*. South Yarra: Hardie Grant Books.

Macdonald, W. (1999) *One Step Beyond*. South Yarra: Hardie Grant Publishing.

Mackintosh-Smith, T. (2001) *Travels with a Tangerine: A Journey in the Footsteps of Ibn Battutah* (2002 edn). London, Basingstoke and Oxford: Picador.

Matrix Shackleton Centenary Expedition (2013) *Henry Worsley: Why I'm Going*. See http://www.shackletoncentenary.org/the-team/why.php (accessed May 2013).

Mear, R. and Swan, R. (1987) *In the Footsteps of Scott*. London: Grafton Books.

Morell, V. (2001) *Blue Nile: Ethiopia's River of Magic and Mystery*. Washington, DC: National Geographic Society, Adventure Press.

Morris, C. (ed.) (1909) *Finding the North Pole* (2003 reprint). Guilford, CT: Lyons Press.

Muir, J. (1894) *The Mountains of California* (1992 reprint). London: Diadem.

Muir, J. (1913) *The Story of My Boyhood and Youth* (1992 reprint). London: Diadem.

Muir, J. (2003) *Alone Across Australia: One Man's Trek Across a Continent*. Camberwell: Penguin Books Australia.

Newby, E. (1958) *A Short Walk in the Hindu Kush* (1981 reprint). London: Pan.

Norvill, P. (1988) *Solo Around the World*. Murrurundi: Peter Norvill.

Olsen, G. (2005) Q & A: Third space tourist. *BBC News*, 4 May. See http://news.bbc.co.uk/go/pr/fr/-/1/hi/sci/tech/3682397.stm (accessed November 2005).

Palin, M. (1989) *Around the World in Eighty Days*. London: BBC Books.

Piccard, B. and Jones, B. (1999) *The Greatest Adventure*. London: Headline Books.

Prescot, C. (2000) *To the Edge of Space: Adventures of a Balloonist*. London: Boxtree.

Quammen, D. (1996) *The Song of the Dodo: Island Biogeography in an Age of Extinction*. London: Pimlico.

Robinson Flannery, N. (2000) *This Everlasting Silence: The Love Letters of Paquita Delprat and Douglas Mawson, 1911–1914* (2005 edn). Melbourne: Melbourne University Press.

Salak, K. (2001) *Four Corners: A Journey into the Heart of Papua New Guinea*. London: Bantam Books.

Scott, R.F. (1905) *Voyage of the Discovery* (1953 reprint). London: Murray.

Selby, B. (1988) *Riding the Desert Trail*. London: Sphere Books.

Severin, T. (1978) *The Brendan Voyage*. New York: McGraw-Hill.

Shekhdar, J. with Griffiths, E. (2001) *Bold Man of the Sea: My Epic Journey*. London: Hodder & Stoughton.

Shuttleworth, M. (2002a) Interview with Mark – 13 April. *First African in Space*. See http://www.firstafricaninspace.com/home/mission/faq/interview5.shtml (accessed November 2005).

Shuttleworth, M. (2002b) On his past. *First African in Space*. See http://www.firstafricaninspace.com/home/mission/faq/leadership.shtml (accessed November 2005).

Shuttleworth, M. (2002c) Interview with Mark – 3 April. *First African in Space*. See http://www.firstafricaninspace.com/home/mission/faq/interview2.shtml (accessed November 2005).

Shuttleworth, M. (2002d) Interview with Mark – 10 April. *First African in Space*. See http://www.firstafricaninspace.com/home/mission/faq/interview4.shtml (accessed November 2005).

Tayler, J. (2003) *Glory in a Camel's Eye: Trekking Through the Moroccan Sahara*. New York: Houghton Mifflin.

Thesiger, W. (1959) *Arabian Sands* (2005 reprint). London: Folio Society.

Thesiger, W. (1964) *The Marsh Arabs* (2005 reprint). London: Folio Society.

Thomson, H. (2001) *The White Rock: An Exploration of the Inca Heartland* (2010 edn). London: Phoenix.

Tschiffely, A.F. (1932) *Southern Cross to Pole Star: Tschiffely's Ride* (1982 reprint). London: Century.

van der Post, L. (1958) *The Lost World of the Kalahari* (2002 reprint). London: Vintage.

Vihlen, H.S. (1971) *April Fool: or How I Sailed from Casablanca to Florida in a Six-Foot Boat.* Chicago, IL: Follett Publishing.

Waterman, J. (2001) *Arctic Crossing: A Journey Through the Northwest Passage and Inuit Culture.* New York: Alfred A. Knopf.

Weihenmayer, E. (2001) *Touch the Top of the World: A Blind Man's Journey to Climb Higher Than the Eye Can See.* Sydney: Hodder Headline.

Wells, K. and Wells, M. (2008) *Camino Footsteps: Reflections on a Journey to Santiago de Compostela.* Fremantle: Fremantle Press.

Wheeler, S. (1992) *Evia: Travels on an Undiscovered Greek Island* (2007 edn). London and New York: Tauris Parke.

Woodhead, P. (2003) *Misadventures in a White Desert.* London: Hodder & Stoughton.

Worsley, H. (2011) *In Shackleton's Footsteps: A Return to the Heart of the Antarctic* (2012 edn). London: Virgin.

Fiction

Adams, D. (1979) *The Hitchhiker's Guide to the Galaxy* (1983 omnibus reprint). New York: Harmony.

Ballantyne, R.M. (1858) *The Coral Island* (1977 reprint). London and New York: Garland.

Barrie, J.M. (1902) *The Admirable Crichton* (1963 reprint). London: University of London Press.

Boulle, P. (1963) *Planet of the Apes* (2011 reprint). London: Vintage.

Christie, A. (1936) *Murder in Mesopotamia* (2002 reprint). London: Harper.

Conrad, J. (1899) *Heart of Darkness* (2007 reprint). London: Vintage.

Defoe, D. (1719) *Robinson Crusoe* (1945 reprint). London: Everyman.

Doyle, A.C. (1912) *The Lost World* (2001 reprint). London: Penguin.

Golding, W. (1954) *Lord of the Flies.* London: Faber & Faber.

Haggard, H.R. (1885) *King Solomon's Mines* (1991 reprint). Oxford and New York: Oxford University Press.

Jerome, J.K. (1889) *Three Men in a Boat* (2011 reprint). London: Vintage.

Martell, Y. (2001) *Life of Pi.* New York: Harcourt.

Morgan, M. (1991) *Mutant Message Down Under* (1995 edn). New York: Harper.

Rice Burroughs, E. (1917) *A Princess of Mars* (2007 reprint). London: Penguin.

Robinson, K.S. (1992) *Red Mars.* New York: Bantam Books.

Robinson, K.S. (1994) *Green Mars.* New York: Bantam Books.

Robinson, K.S. (1996) *Blue Mars.* New York: Bantam Books.

Verne, J. (1862) *Five Weeks in a Balloon* (1996 reprint). Ware: Wordsworth.

Verne, J. (1864) *Journey to the Centre of the Earth* (1996 reprint). Ware: Wordsworth.

Verne, J. (1867–1968) *In Search of the Castaways* (2000 reprint). Project Gutenberg. See http://www.gutenberg.org.

Verne, J. (1870) *Twenty Thousand Leagues Under the Sea* (1993 reprint). London: Everyman.

Verne, J. (1873) *Around the World in Eighty Days* (1996 reprint). Ware: Wordsworth.

Wells, H.G. (1898) *The War of the Worlds* (1946 reprint). Harmondsworth and New York: Penguin.
Wolfe, T. (1979) *The Right Stuff* (1991 reprint). New York: Farrar, Straus & Giroux.
Wyss, J.D. (1812) *The Swiss Family Robinson* (1991 reprint). Oxford and New York: Oxford University Press.

Secondary References

Adler, J. (1989) Travel as performed art. *American Journal of Sociology* 94 (6), 1366–1391.

African Affairs (1964) Major Gordon Laing. *African Affairs* 63 (253), 267–269.

Akerman, J. (1993) Blazing a well-worn path: Cartographic commercialism, highway promotion and automobile tourism in the United States 1880–1930. *Cartographica* 30 (1), 10–20.

Allen, B. (2002) *The Faber Book of Exploration: An Anthology of Worlds Revealed by Explorers Through the Ages.* London: Faber & Faber.

Ambrose, S.E. (1996) *Undaunted Courage: Meriwether Lewis, Thomas Jefferson and the Opening of the American West.* New York: Touchstone.

Anderson, M.J. and Shaw, R.N. (1999) A comparative evaluation of qualitative data analytic techniques in identifying volunteer motivation in tourism. *Tourism Management* 20 (1), 99–106.

Arnould, E. and Price, L. (1993) River magic: Extraordinary experience and the extended service encounter. *Journal of Consumer Research* 20 (1), 24–45.

Ashley, M. (2011) *Out of This World: Science Fiction But Not As You Know It.* London: British Library.

Beattie, D.A. (2001) *Taking Science to the Moon: Lunar Experiments and the Apollo Program.* Baltimore, MA: John Hopkins University Press.

Beedie, P. and Hudson, S. (2003) Emergence of mountain-based adventure tourism. *Annals of Tourism Research* 30 (3), 625–643.

Beinart, W. and Hughes, L. (2007) *Environment and Empire.* Oxford: Oxford University Press.

Belk, R.W. (1992) Moving possessions: An analysis based on personal documents from the 1847–1869 Mormon migration. *Journal of Consumer Research* 19, 339–361.

Belk, R.W. and Costa, J.A. (1998) The mountain man myth: A contemporary consuming fantasy. *Journal of Consumer Research* 25, 218–240.

Blunt, A. (1994) *Travel, Gender and Imperialism: Mary Kingsley and West Africa.* New York and London: Guilford Press.

Bonington, C. (2000) *Quest for Adventure: Remarkable Feats of Exploration and Adventure 1950–2000.* London: Weidenfeld & Nicolson.

Bonyhady, T. (1991) *Burke and Wills: From Melbourne to Myth.* Sydney: David Ell.

Boorstin, D. (1960) Introduction. In I.L. Bird (1879) *A Lady's Life in the Rocky Mountains* (1969 reprint). Norman, OH: University of Oklahoma Press.

Branson, R. (2002) *Losing My Virginity: The Autobiography.* Milsons Point, NSW: Random House.

Branson, R. (2004) Just a big kid. *The Age*, Adventure Special Travel Supplement, 24 July, p. 3.

Bruce, D. (2013) 'Terra Nova' arrival in 1913 remembered. *Otago Daily Times*, 11 February, p. 13.

Butcher, W. (1990) *Verne's Journey to the Centre of the Self: Space and Time in the Voyages Extraordinaires*. Basingstoke: MacMillan.

Butler, R.W. (1990) The role of the media in influencing the choice of vacation destinations. *Tourism Recreation Research* 15 (2), 46–53.

Butler, R.W. (1996) The development of tourism in frontier regions: Issues and approaches. In Y. Gradus and H. Lithwick (eds) *Frontiers in Regional Development* (pp. 213–229). Lanham, MD: Rowman and Littlefield.

Cameron, K.M. (1990) *Into Africa: The Story of the East African Safari*. London: Constable.

Campbell, J. (1949) *The Hero with a Thousand Faces* (1993 reprint). London: Fontana Press.

Carruthers, J. (2009) Full of rubberneck waggons and tourists: The development of tourism in South Africa's national parks and protected areas. In W. Frost and C.M. Hall (eds) *Tourism and National Parks: International Perspectives on Development, Histories and Change* (pp. 238–256). London and New York: Routledge.

Cater, C. (2010) Steps to space: Opportunities for astrotourism. *Tourism Management* 31 (6), 838–845.

Cater, C. and Cloke, P. (2007) Bodies in action: The performativity of adventure tourism. *Anthropology Today* 23 (6), 13–16.

Celsi, R.L., Rose, R.L. and Leigh, T.W. (1993) An exploration of high-risk leisure consumption through skydiving. *Journal of Consumer Research* 20 (1), 1–23.

Chang, K. (2011) Microsoft billionaire aims for the stars. *The Age*, 15 December, p. 14.

Chang, K. (2013) Blow-up space modules expected to take off. *The Saturday Age*, 19 January, p. 11.

Cholidis, N. (1999) 'The glamour of the East': Some reflections on Agatha Christie's *Murder in Mesopotamia*. In C. Trümpler (ed.) *Agatha Christie and Archaeology* (2002 reprint) (pp. 335–349). London: British Museum Press.

Christie, A. (1977) *An Autobiography* (2010 reprint). London: Harper.

Cockell, C. (2007) *Space on Earth: Saving Our World by Seeking Others*. Basingstoke and New York: Macmillan.

Cohen, E. (2004) *Contemporary Tourism: Diversity and Change*. Kidlington: Elsevier.

Collis, C. (2004) The Proclamation Island moment: Making Antarctica Australian. *Law Text Culture* 8, 1–18.

Conefrey, M. and Jordan, T. (1998) *Icemen: A History of the Arctic and Its Explorers*. London: Boxtree.

Conrad, J. (1928) Travel. In *Last Essays* (1955 edn) (pp. 84–92). London: Dent.

Cooper, A. (2012) *Patrick Leigh Fermor: An Adventure*. London: Hodder and Stoughton.

Crane, D. (2005) *Scott of the Antarctic*. New York: Knopf.

Cronin, M. (2000) *Across the Lines: Travel, Language, Translation*. Cork: Cork University Press.

Crouch, G. (2013) *Homo sapiens* on vacation: What can we learn from Darwin? *Journal of Travel Research* 52 (5), 575–590.

Crouch, G.I. and Laing, J.H. (2004) Australian public interest in space tourism and a cross-cultural comparison. *Journal of Tourism Studies* 15 (2), 26–36.

Crouch, G.I., Devinney, T.M., Louviere, J.J. and Islam, T. (2009) Modelling consumer choice behaviour in space tourism. *Tourism Management* 30 (3), 441–454.

Csikszentmihalyi, M. (1975) *Beyond Boredom and Anxiety*. San Francisco, CA: Jossey-Bass.

Cunningham, C. (1996) *The Blue Mountains Rediscovered: Beyond the Myths of Early Australian Exploration*. Sydney: Kangaroo.

Dann, G. (1999) Writing out the tourist in space and time. *Annals of Tourism Research* 26, 159–187.

Davis, W. (2011) *Into the Silence: The Great War, Mallory and the Conquest of Everest*. London: Bodley Head.

Dick, S.J. (1998) *Life on Other Worlds: The 20th-Century Extraterrestrial Life Debate*. Cambridge and New York: Cambridge University Press.

Digance, J. (2003) Pilgrimage at contested sites. *Annals of Tourism Research* 30 (1), 143–159.

Dobbs, D. (2013) Restless genes. *National Geographic*, January, 44–57.

Driver, F. (2001) *Geography Militant: Cultures of Exploration and Empire*. Oxford: Blackwell.

Dunn, R.E. (1986) *The Adventures of Ibn Battuta: A Muslim Traveler of the 14th Century*. Berkeley and Los Angeles, CA: University of California Press.

Edensor, T. (2000) Staging tourism: Tourists as performers. *Annals of Tourism Research* 27 (2), 322–344.

Elliffe, S. (2013) Letters to the editor: Scott100 event did us proud. *Oamaru Mail*, 11 February, p. A6.

Falconer, R. (2005) *Hell in Contemporary Literature: Western Descent Narratives Since 1945*. Edinburgh: Edinburgh University Press.

Feifer, M. (1985) *Going Places*. London: Macmillan.

Florida, R. (2002) *The Rise of the Creative Class: And How It's Transforming Work, Leisure, Community and Everyday Life*. New York: Basic Books.

Forsdick, C. (2005) *Travel in Twentieth-Century French and Francophone Cultures: The Persistence of Diversity*. Oxford: Oxford University Press.

Foster, G.M. (1986) South Seas cruises: A case study of a short-lived society. *Annals of Tourism Research* 13, 215–238.

Frank, K. (1986) *A Voyager Out: The Life of Mary Kingsley* (2005 edn). London and New York: Taurus Parke.

Frost, W. (2004) *Travel and Tour Management*. Sydney: Pearson.

Frost, W. (2010) Life changing experiences: Film and tourists in the Australian Outback. *Annals of Tourism Research* 37 (3), 707–726.

Frost, W. and Hall, C.M. (2009) American invention to international concept: The spread and evolution of national parks. In W. Frost and C.M. Hall (eds) *Tourism and National Parks: International Perspectives on Development, Histories and Change* (pp. 30–44). London and New York: Routledge.

Frost, W. and Laing, J. (2011) Up close and personal: Rethinking zoos and the experience economy. In W. Frost (ed.) *Zoos and Tourism: Conservation, Education, Entertainment?* (pp. 133–142). Bristol: Channel View Publications.

Frost, W. and Laing, J. (2012) Travel as hell: Exploring the *katabatic* structure of travel fiction. *Literature and Aesthetics* 22 (1), 215–233.

Frost, W. and Laing, J. (2013) *Commemorative Events: Memory, Identities, Conflicts*. London and New York: Routledge.

Frost, W. and Laing, J. (2014) The role of fashion in subculture events: Exploring steampunk events. In K. Williams, J. Laing and W. Frost (eds) *Fashion, Design and Events* (pp. 177–190). London: Routledge.

Frost, W. and Laing, J. (forthcoming) Gender, ritual and re-enacting the Wild West: Helldorado Days, Tombstone, Arizona. In J. Laing and W. Frost (eds) *Rituals and Traditional Events in a Modern World*. London: Routledge.

Frost, W., Laing, J., Wheeler, F. and Reeves, K. (2010) Coffee, culture, heritage and destination image: Australia and the Italian model. In L. Joliffe (ed.) *Coffee Culture, Destinations and Tourism* (pp. 99–110). Bristol: Channel View Publications.

Gadney, R. (1983) *Kennedy*. New York: Holt, Rinehart and Winston.

Geiger, J. (2009) *The Third Man Factor: The Secret to Survival in Extreme Environments*. Melbourne: Text.

Glenn, J., Carpenter, S., Shepard, A., Grissom, V., Cooper, G., Slayton, D. and Schirra, W. (1962) *Into Orbit*. London: Cassell.

González, R. and Medina, J. (2003) Cultural tourism and urban management in northwestern Spain: The pilgrimage to Santiago de Compostela. *Tourism Geographies* 5 (4), 446–460.

Gordon, R.J., Brown, A.K. and Bell, J.A. (2013) Expeditions, their films and histories: An introduction. In J.A. Bell, A.K. Brown and R.J. Gordon (eds) *Recreating First Contact: Expeditions, Anthropology and Popular Culture* (pp. 1–30). Washington, DC: Smithsonian Institution Scholarly Press.

Graburn, N. (1983) The anthropology of tourism. *Annals of Tourism Research* 10 (1), 9–33.

Gyimóthy, S. and Mykletun, R.J. (2004) Play in adventure tourism: The case of Arctic trekking. *Annals of Tourism Research* 31 (4), 855–878.

Hall, C.M. (2000) Tourism, national parks and Aboriginal peoples. In R.W. Butler and S.W. Boyd (eds) *Tourism and National Parks: Issues and Implications* (pp. 57–71). Chichester: Wiley.

Hall, C.M. (2002) The changing cultural geography of the frontier: National parks and wilderness as frontier remnant. In S. Krakover and Y. Gradus (eds) *Tourism in Frontier Areas* (pp. 283–298). Lanham, MD: Lexington Books.

Hardesty, D.L. (2003) Mining rushes and landscape learning in the modern world. In M. Rockman and J. Steele (eds) *Colonization of Unfamiliar Landscapes: The Archaeology of Adaptation* (pp. 81–95). Routledge: London.

Hardyment, C. (2000) *Literary Trails: Writers in their Landscapes*. London: National Trust.

Harris, C. and Wilson, E. (2007) Travelling beyond the boundaries of constraint: Women, travel and empowerment. In A. Pritchard, N. Morgan, I. Ateljevic and C. Harris (eds) *Tourism and Gender: Embodiment, Sensuality and Experience* (pp. 235–249). Wallingford: CABI.

Haynes, R.D. (1998) *Seeking the Centre: The Australian Desert in Literature, Art and Film*. Cambridge: Cambridge University Press.

Hiatt, L.R. (1997) Mutant message down under: A new age for old people. In H. Bolitho and C. Wallace-Crabbe (eds) *Approaching Australia: Papers from the Harvard Australian Studies Symposium* (pp. 63–74). Cambridge, MA: Harvard University Press.

Hodges, F. (2007) Language planning and place naming in Australia. *Current Issues in Language Planning* 8 (3), 383–403.

Holland, P. and Huggan, G. (2000) *Tourists with Typewriters: Critical Reflections on Contemporary Travel Writing*. Ann Arbor, MI: University of Michigan Press.

Holtsmark, E.B. (2001) The *katabasis* theme in modern cinema. In M.M. Winkler (ed.) *Classical Myth and Culture in the Cinema* (pp. 23–50). Oxford and New York: Oxford University Press.

Hom Cary, S. (2004) The tourist moment. *Annals of Tourism Research* 31 (1), 61–77.

Howard, C. (2012) Horizons of possibilities: The *Telos* of contemporary Himalayan travel. *Literature & Aesthetics* 22 (1), 131–155.

Hulme, P. and Youngs, T. (2002) Introduction. In P. Hulme and T. Youngs (eds) *The Cambridge Companion to Travel Writing* (pp. 1–13). Cambridge: Cambridge University Press.

Hunt, T. and Lipo, C. (2011) *The Statues that Walked: Unraveling the Mystery of Easter Island.* New York: Free Press.

Hurles, M.E., Maund, E., Nicholson, J., Bosch, E., Renfrew, C., Sykes, B.C. and Jobling, M.A. (2003) Native American Y chromosomes in Polynesia: The genetic impact of the Polynesian slave trade. *American Journal of Human Genetics* 72 (5), 1282–1287.

Ingold, T. (2004) Culture on the ground: The world perceived through the feet. *Journal of Material Culture* 9 (3), 315–340.

Jago, L.K. and Deery, M.A. (2001) Managing volunteers. In S. Drummond and I. Yeoman (eds) *Quality Issues in Heritage Visitor Attractions* (pp. 194–217). Oxford: Butterworth Heinemann.

James, E. (1999) Per ardua ad astra: Authorial choice and the narrative of interstellar travel. In J. Elsner and J. Rubiés (eds) *Voyages & Visions: Towards a Cultural History of Travel* (pp. 252–271). London: Reaktion.

Jeal, T. (1973) *Livingstone* (2013 edn). New Haven, CT: Yale University Press.

Jeal, T. (2007) *Stanley: The Impossible Life of Africa's Greatest Explorer.* London: Faber.

Jeal, T. (2011) *Explorers of the Nile: The Triumph and Tragedy of a Great Victorian Adventure.* London: Faber.

Kane, M. (2010) New Zealand's adventure culture: Is Hillary's legacy a bungy jump? *Annals of Leisure Research* 13 (4), 590–612.

Kane, M. and Tucker, H. (2004) Adventure tourism: The freedom to play with reality. *Tourist Studies* 4 (3), 217–234.

Klint, K.A. (1999) New directions for inquiry into self-concept and adventure experiences. In J.C. Miles and S. Priest (eds) *Adventure Programming* (pp. 163–168). State College, PA: Venture.

Kozinets, R.V. (2001) Utopian enterprise: Articulating the meanings of Star Trek's culture of consumption. *Journal of Consumer Research* 28, 67–88.

Kryza, F. (2006) *The Race for Timbuktu: In Search of Africa's City of Gold.* London: Harper Collins.

Laing, J. (2006) Extraordinary journeys: Motivations behind frontier travel experiences and implications for tourism marketing. Unpublished dissertation, La Trobe University, Australia.

Laing, J.H. and Crouch, G.I. (2004) Flight of fancy: Vacationing in space. In T.V. Singh (ed.) *Novelty Tourism: Strange Experiences and Stranger Practices* (pp. 11–25). Wallingford: CABI.

Laing, J.H. and Crouch, G.I. (2005) Extraordinary journeys: An exploratory cross-cultural study of tourists on the frontier. *Journal of Vacation Marketing* 11 (3), 209–223.

Laing, J.H. and Crouch, G.I. (2009a) Exploring the role of the media in shaping motivations behind frontier travel experiences. *Tourism Analysis* 14 (2), 187–198.

Laing, J.H. and Crouch, G.I. (2009b) Isolation and solitude within the frontier travel experience. *Human Geography/Geografiska Annaler B* 91 (4), 325–342.

Laing, J.H. and Crouch, G.I. (2009c) Myth, adventure and fantasy at the frontier: Metaphors and imagery behind an extraordinary travel experience. *International Journal of Tourism Research* 11 (2), 127–141.

Laing, J.H. and Crouch, G.I. (2011) Frontier tourism: Retracing mythic journeys. *Annals of Tourism Research* 38 (4), 1516–1534.

Laing, J. and Frost, W. (2012) *Books and Travel: Inspiration, Quests and Transformation.* Bristol: Channel View.

Laing, J. and Frost, W. (2015, forthcoming) The new food explorer: Beyond the experience economy. In I. Yeoman, U. McMahon-Beattie, K. Fields, J. Albrecht and K. Meethan (eds) *The Future of Food Tourism.* Bristol: Channel View Publications.

Law, L., Bunnell, L. and Ong, C.-E. (2007) The Beach, the gaze and film tourism. *Tourist Studies* 7, 141–164.

Leed, E. (1991) *The Mind of the Traveler: From Gilgamesh to Global Tourism.* New York: Basic Books.

Leed, E. (1995) *Shores of Discovery: How Expeditionaries Have Constructed the World.* New York: Basic Books.

Lengkeek, J. (2002) A love affair with elsewhere: Love as a metaphor and paradigm for tourist longing. In G. Dann (ed.) *The Tourist as a Metaphor of the Social World* (pp. 189–208). Wallingford: CABI.

Lewis, D. and Bridger, D. (2000) *The Soul of the New Consumer.* London: Nicholas Brealey.

Lyng, S. (1990) Edgework: A social psychological analysis of voluntary risk taking. *American Journal of Sociology* 95 (4), 851–886.

MacCannell, D. (1976) *The Tourist: A New Theory of the Leisure Class.* New York: Schocken.

Mackellar, J. (2006) Fanatics, fans or just good fun? Travel behaviours and motivations of the fanatic. *Journal of Vacation Marketing* 12 (3), 195–217.

Mallowan, M. (1977) *Mallowan's Memoirs.* London: Collins.

Månsson, M. (2011) Mediatized tourism. *Annals of Tourism Research* 38 (4), 1634–1652.

Marin, L. (1993) Frontiers of utopia: Past and present. *Critical Inquiry* 19 (3), 397–420.

Martin, A. (1990) *The Mask of the Prophet: The Extraordinary Fictions of Jules Verne.* Oxford: Clarendon.

May, A. (2006) Two of us: Wilson da Silva & Alan Finkel. Good Weekend, *The Age,* 8 April, p. 18.

Mayne, A. (2003) *Hill End: An Historic Australian Goldfields Landscape.* Melbourne: Melbourne University Press.

McGehee, N.G., Loker-Murphy, L. and Uysal, M. (1996) The Australian international pleasure market: Motivations from a gendered perspective. *Journal of Tourism Studies* 7 (1), 45–56.

McGowan, B. (2006) *Fool's Gold: Myths and Legends of Gold Seeking in Australia.* Sydney: Lothian.

McKie, R. (2012) Have spacesuit, will travel. *Weekend Australian,* 30 June–1 July, pp. 19–21.

Morgan, J. (1984) *Agatha Christie: A Biography.* London: Collins.

Morkham, B. and Staiff, R. (2002) The cinematic tourist. In G. Dann (ed.) *The Tourist as a Metaphor of the Social World* (pp. 297–316). Wallingford: CABI.

Murgatroyd, S. (2002) *The Dig Tree: The Story of Burke and Wills.* Melbourne: Text.

Murphy, G. (2010) *Mars: A Survival Guide.* Pymble: Harper Collins.

Murray, N. (2008) *A Corkscrew is Most Useful: The Travellers of Empire* (2009 edn). London: Abacus.

Nadal, M. (1994) William Golding's *Rites Of Passage:* A case of transtextuality. *Miscelánea: A Journal of English and American Studies* 15, 405–420.

Nepal, S. (2000) Tourism, national parks and local communities. In R.W. Butler and S.W. Boyd (eds) *Tourism and National Parks: Issues and Implications* (pp. 73–94). Chichester: Wiley.

Ness, S. (2012) Talking to my left foot: Performative moves in-between self and landscape in Yosemite National Park. *About Performance* 11, 119–141.

Nevett, T.R. (2004) Marco Polo: International marketing pioneer. *Journal of Macromarketing* 24 (2), 178–185.

Norman, A. (2009) The unexpected real: Negotiating fantasy and reality on the road to Santiago. *Literature and Aesthetics* 19 (2), 50–71.

O'Dell, T. (2005) Experiencescapes: Blurring borders and testing connections. In T. O'Dell and P. Billing (eds) *Experiencescapes: Tourism, Culture and Economy* (pp. 11–33). Copenhagen: Copenhagen Business School Press.

O'Neill, P. (2009) Destination as destiny: Amelia B. Edwards's travel writing. *Frontiers: A Journal of Women Studies* 30 (2), 43–71.

Ooi, C.-S. (2005) A theory of tourist experiences: The management of attention. In T. O'Dell and P. Billing (eds) *Experiencescapes: Tourism, Culture and Economy* (pp. 51–68). Copenhagen: Copenhagen Business School Press.

Ooi, N. and Laing, J.H. (2010) Backpacker tourism: Sustainable and purposeful? Investigating the overlap between backpacker tourism and volunteer tourism motivations. *Journal of Sustainable Tourism* 18 (2), 191–206.

Orion Expedition Cruises (2012) Wanted: 100 people for an expedition. *The Weekend Australian*, Travel, 11–12 February, p. 6.

Paddle, R. (2000) *The Last Tasmanian Tiger: The History and Extinction of the Thylacine.* Oakleigh: Cambridge University Press.

Paradis, T.W. (2002) The political economy of theme development in small urban places: The case of Roswell, New Mexico. *Tourism Geographies* 4 (1), 22–43.

Parkes, C. (2009) *Treasure Island* and the romance of the British Civil Service. In H. Montgomery and N.J. Watson (eds) *Children's Literature: Classic Texts and Contemporary Trends* (pp. 69–80). Basingstoke: Palgrave Macmillan.

Pearce, D.G. (1979) Toward a geography of tourism. *Annals of Tourism Research* 6 (3), 245–272.

Phillips, R. (1999) Writing travel and mapping sexuality: Richard Burton's sotadic zone. In J. Duncan and D. Gregory (eds) *Writes of Passage: Reading Travel Writing* (pp. 70–91). London: Routledge.

Pine, B.J. and Gilmore, J.H. (1999) *The Experience Economy: Work is Theatre & Every Business a Stage.* Boston, MA: Harvard Business School Press.

Pine, B.J. II. and Gilmore, J.H. (2011) *The Experience Economy.* Boston, MA: Harvard Business School Press.

Polezzi, L. (2001) *Translating Travel: Contemporary Italian Travel Writing in English Translation.* Aldershot and Burlington: Ashgate.

Polezzi, L. (2006) Translation, travel, migration. *The Translator* 12 (2), 169–188.

Potts, R. (2008) *Marco Polo Didn't Go There: Stories and Revelations From One Decade as a Post-Modern Travel Writer.* Palo Alto, CA: Solas House.

Poudel, S., Nyaupane, G.P. and Timothy, D.J. (2013) Assessing visitor preference of various roles of tour guides in the Himalayas. *Tourism Analysis* 18 (1), 45–49.

Pratt, M.-L. (2008) *Imperial Eyes: Travel Writing and Transculturation* (2nd edn). New York and Abingdon: Routledge.

Prigg, M. (2013) Thousands of astronauts enter race to take part in '1,000-day mission to Mars' as NASA says red planet is 'top priority'. *Mail Online*, 6 May. See http://

www.dailymail.co.uk/sciencetech/article-2320344/Charles-Bolden-NASA-chief-says-manned-mission-Mars-priority--happen-2033.html (accessed May 2013).

Quan, S. and Wang, N. (2004) Towards a structural model of the tourist experience: An illustration from food experiences in tourism. *Tourism Management* 25, 297–305.

Reddy, M.V., Nica, M. and Wilkes, K. (2012) Space tourism: Research recommendations for the future of the industry and perspectives of potential participants. *Tourism Management* 33 (5), 1093–1102.

Reeves, K. and McConville, C. (2011) Cultural landscape and goldfield heritage: Towards a land management framework for the historic South-West Pacific gold mining landscapes. *Landscape Research* 36 (2), 191–207.

Richards, G. (2001) The experience industry and the creation of attractions. In G. Richards (ed.) *Cultural Attractions and European Tourism* (pp. 55–69). Oxford: CABI.

Richards, G. and Wilson, J. (2004) Travel writers and writers who travel: Nomadic icons for the backpacker subculture? *Journal of Tourism and Cultural Change* 2 (1), 46–68.

Riley, R.W. and Van Doren, C.S. (1992) Movies as tourism promotion. A 'pull' factor in a 'push' location. *Tourism Management* 13 (3), 267–274.

Risse, M. (1998) White knee socks versus photojournalist vests: Distinguishing between travelers and tourists. In C.T. Williams (ed.) *Travel Culture: Essays on What Makes Us Go* (pp. 41–50). Westport, CT and London: Praeger.

Ritchie, B. and Hudson, S. (2009) Understanding and meeting the challenges of consumer/tourist experience research. *International Journal of Tourism Research* 11, 111–126.

Roberts, D. (2013) NASA chief Bolden says agency can go where no man has gone before: Mars. *The Guardian*, 7 May. See http://www.guardian.co.uk/science/2013/may/06/nasa-manned-mission-mars (accessed May 2013).

Robertson, J. and Darby, A. (2013) Explorers take on Antarctic in Shackleton's tracks. *The Age*, 3 January, p. 1.

Robinson, M.E. (2002) Between and beyond the pages: Literature–tourism relationships. In M. Robinson and H.-C. Andersen (eds) *Literature and Tourism: Essays in the Reading and Writing of Tourism* (pp. 39–79). London: Thomson.

Rojek, C. (1993) *Ways of Escape: Modern Transformations in Leisure and Travel*. London: Macmillan.

Ryan, C. (2003) Risk acceptance in adventure tourism – paradox and context. In J. Wilks and S. Page (eds) *Managing Tourist Health and Safety in the New Millennium* (pp. 55–65). Oxford: Pergamon.

Ryan, C. and Bates, C. (1995) A rose by any other name: The motivations of those opening their gardens for a festival. *Festival Management and Event Tourism* 3, 59–71.

Ryan, R. (2013) Emotional, epic ride ends for descendants. *Oamaru Mail*, 11 February, pp. A1–A2.

Sage, V. (2009) Encountering the wilderness, encountering the mist: Nature, romanticism and contemporary paganism. *Anthropology of Consciousness* 20 (1), 27–52.

Sankaran, C. (2008) Narrating to survive: Ethics and aesthetics in Githa Hariharan's *When Dreams Travel*. *Asiatic* 2 (2), 65–72.

Sartre, J.-P. (1938) *Nausea* (trans. R. Baldick) (1973 reprint). Harmondsworth: Penguin.

Saunders, R., Laing, J. and Weiler, B. (2014) Personal transformation through long-distance walking. In S. Filep and P.L. Pearce (eds) *Tourist Experience and Fulfilment: Insights from Positive Psychology* (pp. 127–146). New York: Routledge.

Schmitt, H.H. (2010) Apollo on Mars: Geologists must explore the Red Planet. *Journal of Cosmology* 12, 3506–3516.

Seaton, A.V. (2002) Tourism as metempsychosis and metensomatosis: The personae of eternal recurrence. In G. Dann (ed.) *The Tourist as a Metaphor of the Social World* (pp. 135–168). Wallingford: CABI.

Seligman, M.E.P. (2011) *Flourish*. Sydney: Random House.

Selwyn, T. (ed.) (1996) *The Tourist Image: Myths and Myth Making in Tourism*. Chichester: Wiley.

Shipman, H.L. (1987) *Space 2000 – Meeting the Challenge of a New Era*. New York: Plenum Press.

Shirley, D. with Morton, D. (1998) *Managing Martians*. New York: Broadway.

Singh, S. and Singh, T.V. (2004) Volunteer tourism: New pilgrimages to the Himalayas. In T.V. Singh (ed.) *Novelty Tourism: Strange Experiences and Stranger Practices* (pp. 181–194). Wallingford: CABI.

Smith, S. (2001) *Moving Lives: 20th Century Women's Travel Writing*. Minneapolis, MN and London: University of Minnesota Press.

Smith, V. L. (2000) Space tourism: The 21st century 'frontier'. *Tourism Recreation Research* 25 (3), 5–15.

Sorensen, T. (1965) *Kennedy*. New York: Smithmark.

Spaceport America (2013) *Preview Bus Tours*. See http://spaceportamerica.com/plan-a-visit/preview-tours/ (accessed May 2013).

Squire, S.J. (1988) Wordsworth and Lake District tourism: Romantic reshaping of landscape. *Canadian Geographer* 32 (3), 237–247.

Stebbins, R.A. (1992) *Amateurs, Professionals and Serious Leisure*. Montreal and Kingston: McGill-Queen's University Press.

Strickland, P. (2012) Do space hotels differ from hotels on Earth? The mystery is solved. *Journal of Hospitality Marketing and Management* 21 (8), 897–990.

Suedfeld, P. and Steel, G.D. (2000) The environmental psychology of capsule habitats. *Annual Review of Psychology* 51, 227–253.

Swarbrooke, J., Beard, C., Leckie, S. and Pomfret, G. (2003) *Adventure Tourism: The New Frontier*. Oxford: Butterworth Heinemann.

Tate, K. (2013) How radiation in space poses a threat to human exploration. *Space.com*, 30 May. See http://www.space.com/21353-space-radiation-mars-mission-threat.html (accessed June 2013).

Tedmanson, S. (2013) Adventurers recreate Shackleton's Southern Ocean crossing. *The Press, Christchurch*, 13 February, p. B3.

Telfer, K. (2010) *Peter Pan's First XI: The Story of J.M. Barrie's Cricket Team*. London: Sceptre.

Tenderini, M. and Shandrick, M. (1997) *The Duke of the Abruzzi: An Explorer's Life*. Seattle: The Mountaineers.

Thompson, C. (2011) *Travel Writing*. London and New York: Routledge.

Trauer, B. (2006) Conceptualizing special interest tourism — frameworks for analysis. *Tourism Management* 27 (2), 183–200.

Trümpler, C. (1999a) 'Le camping begins': Life on an archaeological site in the 1930s. In C. Trümpler (ed.) *Agatha Christie and Archaeology* (2002 reprint) (pp. 163–204). London: British Museum Press.

Trümpler, C. (1999b) 'A dark room has been allotted to me ...': Photography and filming by Agatha Christie on the excavation sites. In C. Trümpler (ed.) *Agatha Christie and Archaeology* (2002 reprint) (pp. 229–257). London: British Museum Press.

Trümpler, C. (1999c) Introduction. In C. Trümpler (ed.) *Agatha Christie and Archaeology* (2002 reprint) (pp. 11–16). London: British Museum Press.

Turner, V. and Turner, E. (1978) *Image and Pilgrimage in Christian Culture: Anthropological Perspectives*. Oxford: Basil Blackwell.

Upe, R. (2001) Peak experience. *The Age*, Travel, 15 December, p. 14.

Urry, J. (2002) *The Tourist Gaze* (2nd edn). London: Sage.

Van Nortwick, T. (1992) *Somewhere I Have Never Travelled: The Second Self and The Hero's Journey in Ancient Epic*. New York: Oxford University Press.

Venbrux, E. (2000) Tales of Tiwiness: Tourism and self-determination in an Australian Aboriginal society. *Pacific Tourism Review* 4 (2–3), 137–147.

Verghis, S. (2013) Angel and beast. *The Weekend Australian*, 12–13 January, pp. 4–5.

Virgin Galactic (2013) *Booking*. See http://www.virgingalactic.com/booking/ (accessed May 2013).

Waitt, G., Lane, R. and Head, L. (2003) The boundaries of nature tourism. *Annals of Tourism Research* 30 (3), 523–545.

Walle, A. (1997) Pursuing risk or insight: Marketing adventures. *Annals of Tourism Research* 24, 265–282.

Walter, M. (1999) *The Search for Life on Mars*. St Leonards, NSW: Allen & Unwin.

Wang, N. (1999) Rethinking authenticity in tourism experience. *Annals of Tourism Research* 26 (2), 349–370.

Ward, F. (2012) Cool for cats. *Top Gear*, January, p. 37.

Wearing, S. (2002) Recentering the self in volunteer tourism. In G. Dann (ed.) *The Tourist as a Metaphor of the Social World* (pp. 237–262). Wallingford: CABI.

Week, The (2012) The space race for tourists. *The Week*, 22 June 22. See http://theweek.com/article/index/229536/the-space-race-for-tourists (accessed May 2013).

Wheeler, S. (2003) Introduction. In A. Cherry-Garrard (1922) *The Worst Journey in the World* (2010 reprint). London: Vintage Books.

Wheeler, S. (2007) Introduction. In M. Kingsley (1897) *Travels in West Africa* (2007 reprint). London: Folio Society.

Whybrow, H. (ed.) (2003) *Tales of the Great Explorers 1800–1900*. New York and London: W.W. Norton.

Wilks, J., Pendergast, D. and Leggat, P. (eds) (2006) *Tourism in Turbulent Times: Towards Safe Experiences for Visitors*. Oxford: Elsevier.

Williams, P. and Soutar, G. (2005) Close to the 'edge': Critical issues for adventure tourism operators. *Asia Pacific Journal of Tourism Research* 10 (3), 247–261.

Wilson, E. (2004) 'A journey of her own?' The impact of constraints on women's solo travel. Unpublished dissertation, Griffith University.

Wilson, E.O. (2000) *Sociobiology: The New Synthesis*. Cambridge, MA: Belknap Press of Harvard University Press.

Wolff, J. (1993) On the road again: Metaphors of travel in cultural criticism. *Cultural Studies* 7 (2), 224–239.

Wood, F. (1996) *Did Marco Polo Go to China?* Boulder, CO: Westview Press.

Wood, J., Hysong, S.J., Lugg, D.J. and Harm, D.L. (2000) Is it really so bad? A comparison of positive and negative experiences in Antarctic winter stations. *Environment and Behaviour* 32 (1), 84–110.

Xacobeo (2010) *The Way of Saint James, Galicia*. Galicia: Regional Ministry of Culture and Tourism, Xunta de Galicia.

Young, T. (2009) Framing experiences of Aboriginal Australia: Guidebooks as mediators in backpacker travel. *Tourism Analysis* 14 (2), 155–164.

Zeppel, H. (2009) National parks as cultural landscapes: Indigenous peoples, conservation and tourism. In W. Frost and C.M. Hall (eds) *Tourism and National Parks: International*

Perspectives on Development, Histories and Change (pp. 259–281). London and New York: Routledge.

Zero G Corporation (2013) *Zero g, the weightless experience*. See http://www.gozerog.com/ (accessed April 2013).

Zubrin, R. with Wagner, R. (1996) *The Case for Mars: The Plan to Settle the Red Planet and Why We Must*. New York: Touchstone.

Zurick, D. (1995) *Errant Journeys*. Austin, TX: University of Texas Press.

Index

Explorer Traveller narratives are listed under author. Fictional works are listed under both author and title.